MANAGING OUR WORLD

APPLYING GIS

MANAGING OUR WORLD

GIS FOR NATURAL RESOURCES

Edited by
Geoff Wade
Matt Artz

Esri Press
REDLANDS | CALIFORNIA

Esri Press, 380 New York Street, Redlands, California 92373-8100

Printed in the United States of America.

ISBN: 9781589486881
Library of Congress Control Number: 2022950801

For purchasing and distribution options (both domestic and international), please visit esripress.esri.com.

On the cover: Photograph by Fokke Baarssen.

CONTENTS

INTRODUCTION

AN INCREASING GLOBAL POPULATION, COMBINED WITH the constant desire for economic growth, is putting unrelenting demands on Earth's natural resources. These demands, ranging from agricultural production and timber harvesting to increased use of energy, metals, and other minerals, cannot be sustained indefinitely. In some cases, these demands degrade the resources and critical ecological systems on which we depend. To conserve Earth's resources for future generations, we must find a more equitable and sustainable balance between economic growth and resource conservation. Toward that end, societies and institutions have begun to take steps toward sustainable and regenerative management. This book presents real-life stories about efforts large and small to use resources more wisely and profitably for a better world.

Finding balance

Today, we are starting to see an improved balance that achieves increased economic value by enabling organizations to work better, faster, and more affordably within the framework of a genuine corporate commitment to sustainable practices and social responsibility. This healthy rebalance is driving new corporate behaviors and creating opportunities for geospatial technologies, as shown through the many examples presented in this book.

Increasing economic profitability

Through innovation, simple economically driven geospatial workflow improvements can increase the efficiency of resource production to meet societal demand. These improvements also reduce waste, save time, improve workforce conditions, and lower business risk—for example, through improved field logistics.

Improving environmental protections

Similarly, businesses are using geospatial systems to monitor and mitigate environmental impacts across natural resource industries. These systems demonstrate the viability of adopting aggressive regenerative practices, such as oil-spill contingency planning, sustainable forest management practices, improved mine reclamation, and organic farming.

Expanding societal benefits

Activist investors and society in general expect diversity and inclusion, community outreach, protection of Indigenous rights, and appropriate labor relations and have set new norms for others to meet.

The energy sector faces the challenges of achieving financial profitability, resource sustainability, and societal benefit. Early indicators show that the sector's aggressive investment in renewables supports these goals. Renewables are increasingly economically viable, environmentally clean, and socially desirable. The corporate embrace of renewables motivates energy companies to rebalance their portfolios and support emissions controls, carbon capture innovations, and commitments to net-zero targets.

The real-life stories in this book present initiatives designed to bring about a more sustainable and regenerative balance in practices such as organic farming, mixed-use sustainable forest management, and more. The book aims to highlight some of these efforts, recognize the work of early adopters, and inspire others to follow.

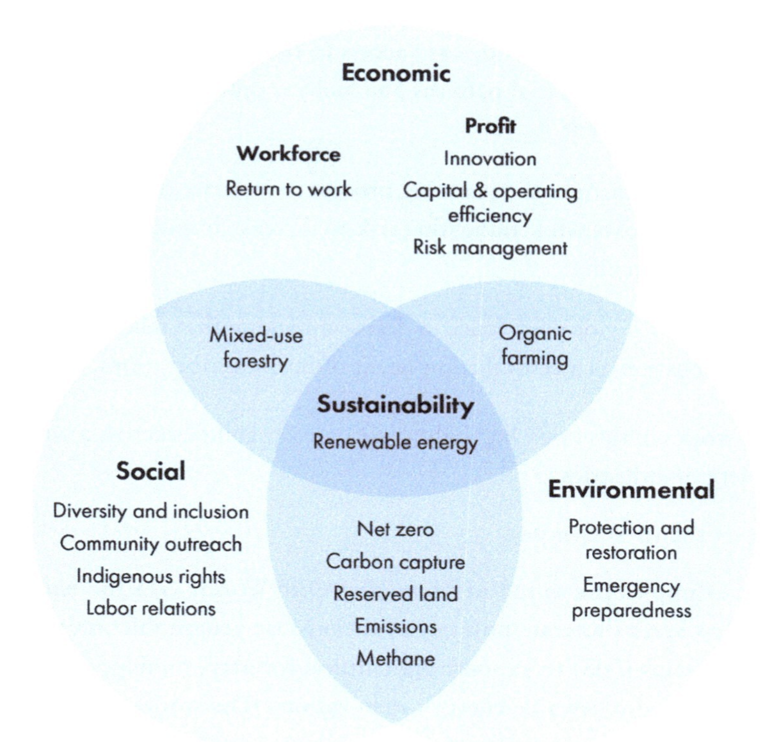

Sustainable use of natural resources requires a balance between economic, social, and environmental considerations.

Workflows

Typical natural resource workflows build on core geospatial system capabilities to visualize, analyze, plan, and act effectively. These workflows benefit users in many ways, helping them to do the following:

- **Visualize:** Bring data to life through a series of maps, apps, and analytics that support an organization's digital transformation journey.

- **Understand:** Provide easy access to relevant information to uncover spatial patterns and analyze options for improvement.
- **Plan:** Energize the planning process to improve operational workflows while mitigating risk to increase business performance.
- **Act:** Provide confidence in decision-making that relies on location as a central component of business operations.

The stories in this book highlight these core capabilities across a wide variety of industries.

Stories and strategies

The examples presented in *Managing Our World: GIS for Natural Resources* illustrate how organizations use geographic information systems (GIS) to support agriculture, forestry, mining, energy, pipeline, and renewable energy organizations. The stories and strategies aim to help you understand GIS and integrate spatial reasoning into natural resource planning and operations using The Geographic Approach. The book concludes with a section about next steps using GIS, which provides ideas, strategies, tools, and suggested actions that you can take to build location intelligence into decision-making and operational workflows.

The Geographic Approach is a way to use geography to help solve problems and make better decisions. If location intelligence isn't currently part of your decision-making processes or considered in an organization's daily operations, or if it isn't used to improve constituent satisfaction, managers can use this book to start developing ideas, approaches, and skills in these areas. Developing these skills does not require GIS expertise, nor does it require you to disregard your previous experience and knowledge. Instead, this book presents location intelligence as a layer of knowledge that managers and practitioners can add to their existing expertise and incorporate successfully into daily operations and planning.

HOW TO USE THIS BOOK

THIS BOOK IS DESIGNED AS A GUIDE TO HELP YOU TAKE first steps with GIS to address issues that are important to you now. Applying The Geographic Approach to decisions and processes can help you solve common problems and create a more collaborative environment in your organization and community. For example, from the many examples in this book, you can identify where maps, spatial analysis, and GIS apps can support your own work and then, as next steps, learn more about how to apply these geospatial capabilities.

Learn more about GIS for natural resources by visiting the web page for this book at:

go.esri.com/mow-resources

PART 1

INTELLIGENT AGRICULTURE

THE USE OF TECHNOLOGY IN THE APPLICATION OF precision agriculture is redefining the science of feeding the planet. New levels of production efficiency, societal responsibility, and nutritional awareness are possible through the application of location intelligence and an increasing awareness and use of geospatial information.

This awareness comes not a moment too soon. Not only do we face the prospect of feeding an estimated 9.5 billion people by 2050, but the world is reckoning with the unforeseen impacts of agriculture on our health and the environment. Impacts include overuse of water in factory farms, increased greenhouse gases caused by deforestation and livestock methane emissions, ocean pollution caused by aquafarming, and marine ecosystems threatened by overfishing. Genetic modification of crops and animals is among additional concerns.

But GIS and related technologies offer promising solutions to meet these and other agricultural and agribusiness challenges of the 21st century. Using GIS, organizations can collect, maintain, analyze, and share agricultural data and make better in-season decisions. Maps, apps, and dashboards can integrate variables, such as soils, irrigation, yield, production costs, profit, and compliance data. Earth observations, imagery, field data, and real-time data streams can help create a digital agricultural or farm twin to increase efficiency and profits while supporting the need for sustainable production to feed the world's growing population.

Next, we'll take a closer look at the use of GIS in farm planning and decision-making, mobile work and connectivity, regenerative agriculture, and public and private partnerships.

Farm planning and decision-making

In response to reduced margins and an uncertain future, agriculture organizations must make planning and decision-making even more efficient to remain competitive. GIS provides these organizations with an integrated digital system for faster decision-making and streamlined operations, aiding them in responding to changes in markets, weather events such as hurricanes, and climate change in general. Organizations benefit from the ability to tune workflows, improve schedules, and model workforce activities in real time, optimizing business performance and ensuring sustainability.

With GIS, agriculture organizations can use data on the locations of crops, workers, vehicles, and processes to establish more efficient workflows and achieve a better balance between profitability and long-term sustainability.

In precision agriculture, the farmer needs detailed information about the field. Key information is gathered from soil testing and yield data to determine the precise amounts of nutrients or seeds to apply. Using precise amounts increases yield and reduces waste, thereby increasing return on investment (ROI). Today, GIS is ubiquitous in precision agriculture. Using spatial information and tracking results year to year, you can meet the goal of precision farming to increase yield and profits in a way that is sustainable for the planet. The use of imagery is also essential in precision agriculture. You can use imagery to detect change through multispectral analysis and understand critical crop status over time.

GIS also helps define planning processes and improve risk management, changing farm management plans from reactive to proactive modeling.

Mobile work and connectivity

Smart agribusiness companies are finding that using GIS to increase efficiency extends to their field operations. Collecting data, often remotely, and analyzing it in near real time with performance dashboards helps organizations model complex workflows, accelerate response times, eliminate waste, and reduce costs. Location technology can help balance and guide the use of natural resources while improving quality and customer satisfaction.

Integrated field apps can be used to collect detailed information at the source, creating a foundation of current field knowledge. With GIS, you can also coordinate, schedule, and dispatch workforce activities. Assigning tasks, monitoring status, and measuring progress in real time increases productivity throughout each step of the supply chain. Collecting and analyzing information in near real-time performance dashboards lets you model complex workflows and improve response times.

Regenerative agriculture

Regenerative agriculture is the practice of balancing profitability and sustainability by protecting the environment, improving soil fertility, and optimizing long-term profitability to create greater food security. Increasingly, precision farmers use geoenabled smart devices and cloud computing to understand how cover crops, rotational grazing, no-till (growing crops with minimal disturbance to the soil), and other sustainable practices contribute to better soil health, biodiversity, and carbon dioxide (CO_2) sequestration.

Bringing location intelligence into regenerative farm management practices leads to new spatial insights, improves decision-making, and creates more balanced outcomes. Collecting information with mobile field apps helps streamline field communications and creates workflow efficiencies. In a sustainable precision agricultural practice, staff can also use imagery and remote sensing data with GIS to detect change through the analysis of multitemporal datasets. GIS lets you explore an array of data types with analytic tools to uncover new patterns and trends that can lead to greater holistic understanding.

Public and private partnerships

Feeding Earth's growing population is a challenge that can't be met without cooperation between the public and private sectors. Food producers must grow more crops with higher nutritional quality, while at the same time minimizing environmental impacts related to soil resources, water consumption, and CO_2 emissions. Producers must also scale sustainable practices, which will require connecting smart farm leadership with policy makers at state and national levels through new business models, to ensure long-term food and economic security.

Field operations must become more efficient, which can be accomplished through automation and digitally supported precision farming. The use of the Internet of Things (IoT) in agriculture will help farmers understand current production status and move toward predictive analytic models.

GIS in action

Next, we'll look at some real-life stories of how organizations are using GIS to meet the agricultural and agribusiness challenges of the 21st century.

THE BUSINESS VALUE OF SUSTAINABILITY

Nespresso

IN OCTOBER 2018, IN QUICK SUCCESSION, THE NOBEL PRIZE in economics was coawarded to a Yale University professor who connected the cost of poor environmental practices to economic health and a United Nations (UN) scientific panel revealed that the world has less than a decade to act against devastating climate change.

The topic of sustainability was once again front-page news.

As governments struggle to find collaborative solutions, businesses are driving corporate sustainability efforts backed by big data analytics and location intelligence.

UN Sustainable Development Goals

On September 25, 2015, heads of state gathered at the UN headquarters in New York City. At that summit, 193 countries agreed to a new set of global sustainability targets. The UN's Sustainable Development Goals (SDGs) use simple language to lay out an international regulatory framework for necessary challenges to be met globally by 2030.

The SDGs are known as a government initiative, but companies worldwide consulted on their development, and these goals could have an enormous impact on the business world. Their implementation will require investment and support from the private sector. A 2015 report by the UN's Sustainable Development Solutions Network estimates the cost of SDGs at $1.4 trillion per year until 2030; in 2020, that estimate was updated to as much as $7 trillion per year. The 2015 report notes that approximately half the investments can be privately financed.

Energy providers, car manufacturers, food purveyors, and other companies that convert natural resources are paying close attention to the sustainability effort. For those businesses, investment in the SDGs could increase the conservation of raw materials of production for decades to come—and result in long-term competitive advantage.

Across the business world, executives can see the implications of supporting SDGs: sustainability could soon become a major business opportunity.

The role of the private sector

One day after the September 2015 summit, the then UN secretary-general Ban Ki-moon held the UN Private Sector Forum to discuss the role of businesses in achieving the SDGs. More than 200 executives from organizations around the world joined him in New York City, including leaders from Dell, Deloitte, Facebook, Fidelity, PepsiCo, and Siemens AG.

"I am counting on the private sector to drive success. Now is the time to mobilize the global business community as never before," the secretary-general told business leaders. "Trillions of dollars in public and private funds are to be redirected towards the SDGs, creating huge opportunities for responsible companies to deliver solutions."

Corporate organizations seem to agree. A 2017 survey by McKinsey Global Surveys found that nearly 60 percent of organizations were more engaged with sustainability than they had been two years earlier, with engagement levels rising to 80 percent in certain industries such as packaged goods and infrastructure.

As climate change continues, companies that rely on natural resources are studying the long-term viability of their products. In some regions of the world, for example, water supplies might soon run out. Reliable cropland could turn fallow as temperatures and weather systems shift. And yet, just 21 percent of business executives told McKinsey that business growth was a top driver of their sustainability initiatives. One way to read that finding is that a select few industry leaders have figured out that smart, sustainable practices can bring about long-term growth and competitive advantage.

Innovative companies are adopting tools such as artificial intelligence (AI), IoT, and analytics to address the SDGs in ways that also benefit the business—doing well by doing good. According to the McKinsey report, nearly half the organizations using technology to advance sustainability are using big data and advanced analytics, which typically include location intelligence.

Where sustainability meets opportunity

One such company is Nespresso. An autonomously managed subsidiary of Nestlé Group, Nespresso is known globally for its premium single-serving coffees. Nespresso's success and customer loyalty result from the company's emphasis on—and investment in—the consistency of its coffee's flavor.

Coffee is a delicate crop, frequently grown in developing countries and dependent on healthy ecosystems. Coffee—and companies such as Nespresso—is susceptible to the increasingly volatile effects of sociocultural dynamics and climate change. For Nespresso, acting today to avoid the perils of tomorrow is not just good citizenship, it's sustainable business.

"Sustainability is really at the core of our business. It is an imperative to our long-term business success," explains Yann De Pietro, operations and sustainability technology manager for coffee at Nespresso. "There have been studies saying that by 2050, Arabica coffee may not be available anymore in some countries if we don't do anything now."

Controlling challenges through sustainability

Nespresso has made a deliberate choice to integrate these challenges into its decision-making process and act on them through sustainability programs. These programs help convert liabilities into business opportunities while supporting the farmers and communities that grow coffee.

Nespresso works with more than 100,000 farmers in 13 countries, up from 300 farmers in 2003. In 2003, the company launched its responsible coffee sourcing program, the Nespresso AAA Sustainable Quality Program, in partnership with the Rainforest Alliance. The program is based on the idea that high-quality coffee and the sustainability of farming communities are interconnected and that building trusting, long-standing relationships with coffee producers will benefit everyone.

The company supports the implementation of sustainable agricultural practices at farms by investing in technical assistance, paying premiums directly to coffee farmers, and cofinancing infrastructure improvements.

The company has invested in a network of more than 450 agronomists—specialists who provide coffee growers with on-site technical assistance and training on practices such as pruning, crop renovation, fair treatment of workers, water usage, and biodiversity conservation, all of which can earn farmers industry certifications.

Through its coffee sourcing program, Nespresso in recent years has invested about $35 million annually in technical assistance and premiums paid to farmers for their quality coffee. The educational program is free to farmers and doesn't require them to sell to Nespresso, De Pietro said. But the benefits help create long-lasting relationships and loyalty.

Nespresso's coffee sourcing program is part of the company's strategic framework called The Positive Cup, which focuses on four areas: coffee, aluminum, climate change, and engagement. Nespresso has committed to milestones that include sourcing 100 percent of its aluminum from responsible sources certified by the Aluminium Stewardship Initiative (ASI), offering consumers convenient solutions for recycling, reducing the carbon footprint, and reaching carbon neutrality for its operations. In 2016, the company tied those efforts to the UN's SDGs, committing to making an impact on 11 of the 17 SDGs.

Progress through digital transformation

Although its sustainability program has been in effect for years, Nespresso has seen rapid results because of advances in digital technology.

At Nespresso, "Digital transformation is a key change for sustainability," De Pietro said. "[We] want to provide maximum impact. So we need the tools to help us to maximize our efforts."

At the center of Nespresso's digital transformation is location intelligence. The company's monitoring and evaluation system uses advanced digital technology that records, maps, and shares data about farms, farmers, and coffee crops. The system reveals local feedback and insight on the impact of its coffee sourcing program and the status of each farm, including its objectives, achievements, and performance. The digital platform—which is powered by GIS and data analytics—reveals insights into the way farmers deliver coffee beans to central mills to be harvested, an important factor in supply chain productivity and efficiency.

Bringing intelligence to location data

One of De Pietro's goals is to help farmers get their crop to market more efficiently. An analysis in Colombia exemplified how location intelligence can create business advantage for the company and its partners.

A location analysis revealed that farmers brought their crops to certain Colombian mills—many of them close to their farms—less frequently than projected. De Pietro queried the GIS technology to study the data further so that he could understand these behavioral patterns. What he discovered was a reminder of topography's effect on time to market.

With basic maps, he said, Nespresso could determine the distance between farmers and mills. But only with sophisticated location intelligence could it fully understand the travel distances to each central mill. The analysis uncovered areas where the terrain required long rides or walks through the mountains to reach certain farms, making frequent visits impractical. De Pietro and his team applied a similar analysis to the travel of agronomists who visit Nespresso's farms and found a similar pattern. For a company that works with 100,000 farmers, having a digital solution to deliver that kind of intelligence is invaluable.

Location intelligence pointed the way to better business and sustainability practices. If the mills were more centrally located, farmers could get coffee to market more quickly. The ability of agronomists to reach farms faster also increased Nespresso's capability to source its coffee from sustainable farmers.

Just as retailers and logistics companies use location intelligence technology to identify and plan the most efficient drive times for customers or delivery workers, Nespresso embraces the idea that the distance to a location is less important than the amount of time it takes a customer or farmer to get there.

Nespresso's use of GIS and location intelligence to build a comprehensive view of farming operations and accessibility across regions is part of its goal of farm accessibility, De Pietro said.

As Nespresso makes progress in Colombia and around the world, the organization maps its efforts back to the UN's SDGs to organize and guide strategic decision-making. The coffee sourcing program, for instance, supports SDGs for inclusive growth, sustainable agriculture, eradication of poverty, water stewardship, and more.

The future of sustainability

The use of location intelligence to reveal the granular details of day-to-day coffee farming benefits Nespresso and farmers. By examining and adjusting locations for farmers, the company frees up precious time and increases productivity. This allows more time for education and strategic planning—the activities Nespresso hopes will sustain its coffee crops for years to come.

Questions remain on ways to implement wide-scale change across industries. The UN reports that, as a whole, its member countries are not on schedule to meet the goals by 2030. "This ambitious agenda necessitates profound change that goes beyond business as usual," the UN has stated.

Business-as-usual attitudes are among sustainability's challenges. Some of the largest and wealthiest organizations in the world are not actively participating in sustainability efforts. The UN's *SDG Commitment Report 100*, released in 2017, found that American companies scored far worse than their European counterparts. Of the 18 companies with no mention of sustainable development themes, 15 were American.

Regardless, there are strong incentives to invest in corporate sustainability strategies, for reasons reactive and proactive.

For instance, a team at McKinsey found that risk-related sustainability issues can impact up to 70 percent of an organization's earnings. Nespresso's core product and the heart of its brand—coffee—is at risk from climate change in the coming years. The company is approaching this challenge proactively, using digital technology and location intelligence to find strategic solutions.

Treating sustainability as a guiding principle and an opportunity to gain competitive value may be the way forward for other innovators in the business community.

A version of this story by Chris Chiappinelli originally appeared in *WhereNext* on December 11, 2018.

USING GIS TO MANAGE STRATEGIC ASSETS

Land O'Lakes

IN THE AGRICULTURE INDUSTRY, EFFICIENCIES AND PROFITS are often influenced by the distance between the farm and various assets—the nearest feedlot, grain elevator, and crop nutrient facility.

More than a century ago, transportation wasn't nearly as efficient as it is today. So, grain elevators were set up in locations that farmers could reach via horse-drawn carts over the course of their workday. Later, these storage facilities were built in locations that could be served by railroads, and trucks began to fill highways to transport grain, feed, and other agricultural commodities. Today, in the digital age, also known as the information age, existing storage facilities are sometimes located in suboptimal areas to serve the farms' current customers, yet new facilities are often built on the same footprints as the ones they replace without consideration of more strategic locations.

Land O'Lakes, an agricultural cooperative with about 4,000 members, streamlines the quality, consistency, marketing, and economics of the farms' dairy products.

The Land O'Lakes agricultural cooperative, based in the Minneapolis–St. Paul suburb of Arden Hills, Minnesota, was founded nearly 100 years ago by 320 creameries to improve the quality, consistency, marketing, and economics of its dairy products. Today, it is one of the largest co-ops in the United States, comprising about 4,000 members and more than 10,000 employees. Its industries include dairy products; animal feed; crop inputs such as fertilizer and water; and insights, including data and guidance. Sales can total up to $15 billion annually.

Several years ago, the co-op formed a strategic asset management (SAM) team, a consulting group within Land O'Lakes that uses GIS technology to serve the company's agriculture retail owners.

The geographic dynamics of business

Land O'Lakes uses GIS for three distinct purposes. First, the SAM team uses geospatial technology for the consulting projects it conducts for co-op members, providing them with insights and solutions for their businesses.

"We show them mapping and locational information about the trade areas for [a] project—where the trade area is, where their facilities are within the trade area, and where the competitors are located," said Josie Taylor, the SAM team's consulting manager until 2021. "This allows us to visually tell them a story about the geographic dynamics of the trade area we are analyzing and how we are thinking about improving it."

Second, the team uses GIS for field data collection. "We often need to go out and look at the facilities that the co-ops are running and analyze how they are operating them. What are the capabilities of those facilities? How old are they? Can we expand those facilities?" Taylor said. "We use [ArcGIS® Survey123] to capture a comprehensive and consistent set of data about the facilities."

Third, Land O'Lakes uses GIS for analytic purposes. As Taylor explained: "If we've got multiple facilities that are close together, we analyze where the trade areas overlap. We also perform transportation analysis to determine how products are currently being delivered to our customers and if there is a more efficient way to do it. That's often about reallocating customers to a specific facility—that is, analyzing the facility they are currently using compared to one that is more appropriate for their use based on distance, functionality, future growth, and so on."

Taylor and her team use GIS in other ways too—for example, to explore drive-time routes within specific trade territories and for route planning. "But that's a little bit more on the fringes of what we do," she said.

Market analyses steer new facility construction

Agriculture is affected by many factors, including changes in strategic assets, such as the consolidation of existing facilities and the construction of new ones. These changes can spontaneously create trade area overlaps, gaps, and redundancies, which inevitably impact efficiency.

A big part of facility rationalization is understanding how much capacity an operation has and how much it needs to serve existing and future customers. Before new facilities are built, the SAM team evaluates each project based on the future needs of the marketplace and the potential ROI. The assessment is based on factors such as the current feed business, customer needs, competitor capabilities, asset efficiency, financial performance, regulatory compliance, market trends, and the risk and exposure of the project.

The group also offers recommendations on how its members can change the look of their transportation or distribution model, Taylor said. "For example, suppose you are currently operating seven feed mills for hog producers. Two of those mills are old, and you don't want to put any more money into them. With some investment, another mill could be operated for an additional five years before you need to replace it. In addition, your market is growing, and you need to increase capacity as the market expands. So, the question is, Where is the best place to put a new facility—to replace the three that are outdated—that can manufacture feed in a way that's more efficient, safer for employees, and implements a tracking mechanism to meet the consumer transparency demands that are needed today?" A transportation study is one of the many measures that the team considers in recommending greater operational efficiencies in the supply chain for a feed manufacturer, Taylor said.

Land O'Lakes also uses transportation studies to examine operating expenses and future transportation costs as part of its financial analysis for capital investment purposes. Such an analysis includes a series of questions:

- How many miles must truck drivers drive to reach distribution facilities?
- How many miles would they need to drive if a new transportation model was implemented to accommodate a different set of distribution facilities?
- Would building these new facilities produce cost savings, or would costs increase?

"Other components of our market analyses include an examination of the trends that are happening in an area in terms of where the demand for our products is coming from," Taylor said. "Going forward, is that demand changing? Who is driving the demand? How much growth has happened in the market historically? Where are

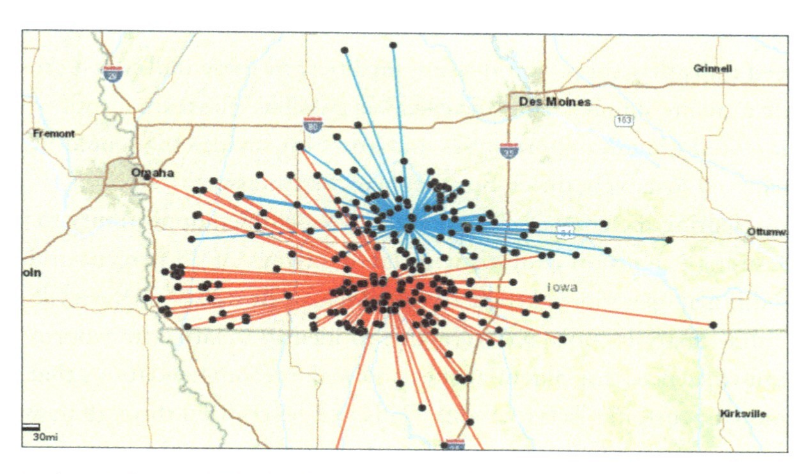

With GIS, the Land O'Lakes SAM team can show co-op members the customer allocations for various facilities—in this case, two mills.

the growth opportunities in the future? Who are the competitors? What are they doing differently than [us], and what is their impact on the market? So we make a recommendation about what should be changed to meet the future needs of an operation."

Adapting to change using GIS

The SAM team also uses GIS for market and logistics analyses when deciding whether co-ops should make large-scale capital investments.

"In our consulting practice, we are seeing increased demand for projects that provide long-term solutions in the changing agricultural industry, and GIS is critical in performing the necessary analyses," Taylor said.

For instance, export demands drive hog production in the United States, so production must be efficient. One of the team's projects—a feed market assessment in Iowa—focused on how to keep up with strong market growth in areas where animal inventories have increased by 5 to 6 percent annually. Sales margins are thin for the feed mills that supply production facilities, so these mills must create as many supply-chain efficiencies as possible. The team identified potential growth opportunities and areas for savings that could be achieved by having the co-ops use their current assets.

"During the course of the project, we identified opportunity for business expansion coupled with labor savings of 23 percent and additional operational benefits," said Taylor. "By using Esri's ArcGIS Online analysis tools, we were able to identify relative transportation efficiencies for potential sites and make recommendations that improve customer serviceability. This will be realized through new facility construction that promotes sharing the current production load among producers and an efficient reallocation of our customers to existing and new facilities."

Continuing to enhance decision-making for long-term planning

The SAM team also faces the challenge of meeting retail owners' business and development needs. GIS helps Land O'Lakes examine how the co-op's members manage their supply and inventory in season and out of season. With this data, the team can help them continually improve their operations.

"A lot of our work isn't about the day-to-day operations of the business. It's about long-term planning," Taylor said. "Our goal is to be the adviser of choice for significant retail agricultural business decisions, whether it's capital or business investments, partnerships, acquisitions, mergers, or achieving the best operational efficiencies."

A version of this story titled "As Agriculture Transforms, Land O'Lakes Uses GIS to Manage Strategic Assets" originally appeared in the Fall 2019 issue of *ArcNews*.

HOW A DATA-DRIVEN COMPANY WINS THE MARKET

John Deere

AS SOME COMPANIES PULL AHEAD, OTHERS FALL BEHIND. It's the essence of competition. Increasingly, data separates the winners from the laggards. Giants such as Amazon, Facebook, and Google have propelled themselves to the forefront of the global economy in large part because of the data they absorb and rapidly transform into operational insights, predictions, new services, and new products. Today, the need for and use of data have spread well beyond Silicon Valley into industries ranging from entertainment, where Netflix's algorithms made it a streaming colossus, to agriculture, where John Deere has revolutionized the agriculture tech space.

Successful executives search for advantage through unique access to metrics about their markets and customers. As Brian Kilcourse, managing partner at RSR Research, put it in a podcast interview, business leaders are seeking to turn this information into something they can use "to clobber the competition."

Market development through the lens of data

Data analysis has countless applications for businesses, from optimizing supply chains and honing marketing messages to creating new product lines by bringing proprietary data to the market. Data analysis yields the best results in creating a more textured, binding relationship with the customer. Companies that outperform competitors use location-based data to understand their customers, where they're located, and how close they are to the things they want or need.

That's why location intelligence has become an essential tool for market development. With a location-centric view of data, companies such as John Deere learn how customers in a given market will likely respond to various products and services—even how much they will spend. Using an AI-assisted GIS, data scientists and analysts at Deere can visualize billions of data points, finding revenue-generating patterns that a human alone could never detect. This kind of predictive view makes it easier than ever to spot unseen expansion opportunities and potential investment traps.

John Deere in a new age of agriculture

In the crowded market for agricultural machinery and heavy equipment, John Deere commands a strong brand awareness, with its recognizable green-and-yellow color scheme and leaping deer logo. Founded in 1837, it has also embraced smart data practices.

"As our CEO recently said, we're a technology company," said Angela Bowman, principal scientist for John Deere. "That's it, first and foremost." Thanks to sophisticated machines equipped with sensors that capture data on soil, water, and temperature conditions, Deere has collected immense amounts of useful information. Satellite imagery lets the company's decision-makers analyze land cover, sizing up how various grasslands, crop fields, or lawns correlate with consumer purchases. A map with 50 billion data points about field conditions and topography gathered from IoT-equipped machines gives the company a kind of intelligent "nervous system" of US farms and lawns. Deere's use of data has caught the attention of people such as Jennifer Belissent, a former principal analyst at the research and advisory firm Forrester which specializes in the data economy.

"John Deere is really illustrating that companies that use data in intelligent and strategic ways are gaining competitive advantage in a lot of different ways and in a lot of different industries," Belissent said.

Forrester surveys have shown that 56 percent of firms are expanding their ability to source external data, and another 20 percent plan to do so in the year ahead.

Deere uses location information to assess potential markets and predict growth opportunities. In essence, the company has turned the art of market assessment into a science, forming action plans based on the location intelligence it distills from sales data, demographics, land-cover and satellite imagery, and competitive insight.

Seeing the future of farming and retail

To forecast in what can be a precarious trade, farmers have always relied on some form of predictive intelligence, from bird flight patterns and cloud formations to the folksy divinations of the *Farmers' Almanac*. As large-scale farming has become more sophisticated and technologically advanced—turning into what is often called precision agriculture—Deere has placed itself at the forefront of predictive intelligence. This includes helping customers forecast crop yields and needed machine maintenance as well as advising dealers on which markets will perform best.

To outsiders, the success of a Deere dealership might seem like a matter of luck, a product of good or bad weather. But Deere uses location data to remove the guesswork from market development, supporting retail dealer investments with a foundation of objective analysis.

"The work I do for market planning would be impossible without location intelligence," said Mattias Wallin, a data scientist with the market research group at John Deere. Using GIS technology, Wallin delivers market intel to Deere's Dealer Development office, whose professionals advise dealers with data-based investment predictions.

For many of these dealers, opening another location would be one of the biggest capital expenditures of their professional lives—a potentially risky bid for growth that they pursue only a couple of times in a career. With location intelligence, Deere can increase the odds of dealer success by making market assessment less about gut instinct and more of a calculated chess move that considers all the variables.

"The whole point is to remove uncertainty," Wallin said. That gives dealers confidence that they have made the right decision in expanding to a location. This approach generated about $44 billion in worldwide net sales and revenue for Deere in 2021, according to the company.

Tools of the data trade

During the early stages of a market analysis, Wallin and his team might sift through thousands of variables related to sales and land data, demographics, and other kinds of information. They typically need as few as 20 of these variables to accurately predict the commercial value of a potential area, usually measured in census blocks. These blocks generally include about 3,000 people and range in size from a city block in urban locations to hundreds of square miles in remote or rural regions.

The cornerstone of Wallin's work is sales data—the hard numbers on what products sold where and when they sold best. Using GIS technology, Wallin can perform an AI-powered regression analysis on a potential new market. The analysis entails looking at other currently active markets with similar characteristics of land cover, customer sales, and demographics and projecting potential revenue for the new area based on past performance. If the market characteristics are different—lawns instead of crop fields or higher or lower income levels—the GIS-based AI model examines how those variables affected revenue in other areas, then folds that into the analysis of the potential site. These insights allow dealers to see opportunities at a granular level, gaining important insight that the competition often lacks.

To demonstrate the power of location intelligence in quantifying markets, Wallin first applies the analysis to the dealer's current region, treating it as if it were a new market and sharing the results with the dealer—often to their surprise. "A dealer in the Pacific Northwest said, 'Wow, Mattias, your tool captured everything that I've learned about my markets and customers for the last 30 years.'"

Psychographics: Insights into customer lifestyle

As with any market analysis, demographic information is essential, but psychographic data—a sophisticated view of consumer behaviors and tastes—can guide precise store siting and even influence the store's product mix. This is an important advantage for a company such as John Deere, whose products can range in cost from $1,600 to $600,000, leading to a widely segmented audience.

"I really see the value in psychographics as a way to visualize or bring some context around 'Who is your customer? What do they look like?'" Wallin said.

A GIS analysis of a market's psychographics might reveal, for example, that an exurban area is home to high-income, white-collar professionals with a taste for the farming lifestyle. These individuals could own a few acres of grass and be willing to spend more money on a high-end lawnmower or small utility tractor.

A John Deere dealer aiming to court this customer segment might use GIS to pinpoint sites with clear traffic lanes to exurban areas and comfortable proximity to direct competitors. The data could also shape which products the dealer stocks and which it makes visible from the road. Online and direct-mail marketing campaigns could be targeted to more affluent consumers in zip codes linked to higher-acreage homes. These are all decisions a data-driven company can make confidently, based on details likely to yield substantial gains in the long run.

Who gets there first?

John Deere's successful use of location intelligence and psychographic information underscores the central insight of analysts such as Belissent—that the most effective ways to build data into a business model often center on the relationship with customers and serving their needs. "The number one thing we're seeing in terms of driving the use of data is improving customer experience," Belissent said.

Companies that crunch data to improve operational efficiency alone tend to progress at a slower rate. Those that pull in external data—whether it's weather patterns, satellite imagery, or demographic and psychographic profiles—tend to better understand the markets they hope to serve and place supply closer to demand. Developing that location intelligence quickly in a fast-paced business environment increasingly depends on savvy use of data.

"If somebody is asking for insight that requires some advanced analytics capability and it's taking months or years, somebody else is going to get there first," Belissent said. "The competitor is going to get there first."

A version of this story by Marianna Kantor and Frits van der Schaaf titled "How Data-Driven John Deere Wins the Market" originally appeared in *WhereNext* on October 8, 2019.

OVERCOMING THE GAP BETWEEN INFORMATION TECHNOLOGY AND BUSINESS USERS

Iowa Select Farms

IT'S A TALE AS OLD AS PUNCH CARDS, AND THOUGH IT HAS received coverage in publications as prominent as the *MIT Sloan Management Review* and *Harvard Business Review*, the schism between information technology (IT) and business managers persists. Business managers often request IT's help creating apps and reports for internal teams, partners, and customers. IT, with its focus on security and compliance, approaches the work cautiously, often not at the pace that business users expect.

Both say they sometimes speak different languages and often misunderstand each other's needs and constraints. But despite these challenges, a leading pork producer in Iowa found a way to bridge the gap.

In early 2019, a business team at Iowa Select Farms sought help from its IT department to use GIS to create more precise, efficient, and environmentally responsible farming. Even with a cooperative IT department, colleagues faced a thorny challenge: everyone involved had grown up around farming, but they all had to learn each other's language.

Ultimately, their work together led to a success story with lessons for any business—not the least of which was that innovation can be found beyond the Fortune 500.

The need for a mobile app to keep the company growing

Iowa Select Farms needed an app powered by GIS for farmers who use the by-product of hog farming to fertilize thousands of acres of cropland in precise and environmentally responsible ways. Hog manure is a valuable organic fertilizer that improves soil health, increases the carbon-storing capacity of soil, and reduces erosion and runoff.

As the fourth-largest pork producer in the United States, the Iowa Select Farms network includes 800 hog farms across 50 Iowa counties, many run by independent farmers and landowners. The company sells 5 million hogs a year that grow to market size in feeding barns around the state—contributing to a $20 billion US industry. The animals' manure is collected in various storage structures, including belowground, environmentally controlled concrete containment facilities.

A group of about 100 companies use manure application equipment to transfer the by-product to the field—a process meant to honor environmental regulations while maximizing crop yield. Knowing where to apply fertilizer demands geographic precision. With paper-based maps stored in three-ring binders, the contractors who drove the tractors weren't as efficient as they could be.

As part of its commitment to continuous improvement, Iowa Select Farms wanted an app that was simple enough for the equipment operator to read on a cell phone. But as Iowa Select Farms teams began to create a mobile app, they found huge amounts of data to manage, strict deadlines to meet, and stringent regulations to follow.

The effort to create an easy-to-use app for drivers of manure application rigs may not represent the most glamorous part of the business world, but it benefits the state's crops and conservation efforts, which in turn provide food for people and feed for hogs.

A look inside business–IT collaboration

Nick Johansen, environmental services manager and GIS coordinator for Iowa Select Farms, asked IT to develop the app. The IT team turned to GIS to analyze how much fertilizer to apply to which portions of the fields in accordance with environmental regulations—a form of location intelligence key to smart agriculture.

At times, the teams involved in building and testing the app—all Iowans with at least some farming background—wondered if they would find common ground between the business-speak on one side and IT-speak on the other.

"We had a fair number of meetings where we were talking our language and they were talking their language, and [we] got out of the meeting and it's like, 'You don't get it,'" said Keith Kratchmer, director of nutrient management and compliance, and Johansen's boss.

Still, when he and Johansen first sat down with IT, they were surprised by the excitement of the developers.

IT director Carl Vogel was surprised, too. "The developer I had tasked with the project was really passionate about GIS data. Loved it. Nick, on the other hand, also loves that, and he also is really passionate about just working with farmers, working in agriculture."

Shared enthusiasm set a foundation, but the Iowa Select Farms teams needed much more to build what would become the aptly named Pumper Portal. It would require focused discussions and, sometimes, faith in the outcome.

"The dynamic that we've built, the rapport we've built, is that we both knew what the goal was, and we knew how it was going to get done," Johansen said. "We had to work on it. We all had no idea how long this was going to take, especially on our end, the user end. We expected [IT] to sprinkle the pixie dust on it and make it happen."

Vogel said the conversations with Johansen and Kratchmer were productive. "We had a good mix of professionals who were all focused on making something useful for our end user. Everybody involved kept a professional attitude, and that helped us meet our goals."

Fertile ground for success

Manure is an important part of the hog farming economy—and the soybean and corn cultivation that often occurs on hog farms.

Kratchmer and Johansen's nutrient management group oversees the effective and environmentally appropriate management of animal manure. They wanted to make the process more sustainable and efficient for 100 or so contractor companies that service farms across the state.

On its website, Iowa Select Farms notes: "We're proud to be a sustainable food system by providing our neighboring crop farmers with manure to replenish their soil's fertility. Full of essential crop nutrients such as nitrogen, potassium, and phosphorus, manure brings pork production full circle by feeding the crops that feed the pigs."

On the farm, operators of application equipment collected manure from the containment structures under the hog enclosure,

An applicator truck prepares to follow its prescribed route through a field. Image courtesy of Iowa Select Farms.

and then injected the nutrients into the ground, allowing them to diffuse through the soil and fertilize crops. The Pumper Portal app created by the IT and nutrient management teams guided each step of the process—a form of precision agriculture that delivers better crop yields while improving soil health.

The app allows the applicator and the Iowa Select Farms team to access data on recent soil analysis, which determines exactly how much fertilizer should be applied to specific zones of the field. The app accounts for topographical features that require special care to ensure minimal runoff and maximum nutrient value to the crop. The engine behind that precision is GIS technology, which tracks ever-changing conditions on a farm just as it tracks weather moving across the country, anonymous shoppers moving through a city, or planes in the sky.

"The Pumper Portal really just allowed us to make our data sharing with our applicators more streamlined," Johansen said. "The applicators can see it on their phones, tablets, computers—however they need to, from wherever. It's very interactive. We serve up a sample of our GIS so they see the field boundaries as we have them mapped."

Everyone who worked on the app knew that apart from the system's sophisticated inner workings, the user interface needed to be easy to navigate. Throughout the process, Kratchmer asked himself, "Are we going to make it simple enough that people will use it?

Respecting the other person's professional language

Building a complicated app from scratch required a shared belief in a common goal—one that helps the worker, the company, and the environment.

During app development, the IT and business teams witnessed the commitment of their colleagues.

"I think you've got to get to know and respect the person that you're going to be working with," Kratchmer said. His advice for building bridges between the groups: "Educate yourself so you can try to bridge that gap between the language IT speaks and the language that you're trying to speak, so you can get to a common goal."

That tactic paid dividends for Iowa Select Farms as the effort led to common understanding, Kratchmer said. That happened in part because Vogel, unlike some IT directors, believes app developers should interact directly with business teams, rather than receive requests through intermediaries. That direct line of communication helped lead to a successful Pumper Portal app.

When the nutrient management crew unveiled Pumper Portal to end users, Vogel sent his developer to see the reaction, answer questions, and seek feedback. The app drew on sophisticated location technology, but the interface was simple and clean, and the applicators—even the tech-averse ones—quickly used it in the field.

Lessons learned, lessons shared

The teams found that the human dimension played a big role in the app's success, with these takeaways:

- **Check in your ego:** "I think that's a part of what we do in our department working with so many different farmers and companies," Johansen said. "It's our job to make it all happen. Your ego can't get in the way of that."

- **Emphasize the team:** "Everybody's got to be willing to work together—that's a big deal," Kratchmer said. He attributes much of the team's success to willingness to learn the language of the other camp and each person's patience in working through constraints.

- **Find the passionate ones:** "It comes down to the people," Vogel said.

Ultimately, the app was only as good as the people who built it and use it, Vogel said. "We got a good alignment of unique individuals who were passionate about this," he said, "and the stars lined up really well for that to be successful."

A version of this story by Mark Dann titled "How One Company Overcame the Gap between IT and Business Users" originally appeared in *WhereNext* on May 20, 2021.

TRACING EACH OYSTER FROM TIDE TO TABLE

Taylor Shellfish Farms

TAYLOR SHELLFISH FARMS, THE LARGEST SHELLFISH producer in the United States, recently added a genetics program. Rising ocean acidity—and the need to improve the resilience of the clams and oysters that Taylor raises—helped drive this decision.

Starting in 2005, the company began to feel the impact of acidity on its 30 farms, which span 10,000 acres of tidelands in the Pacific Northwest. Hatcheries and wild oysters across the region experienced a dramatic die-off estimated in the billions. By 2008, production at its hatcheries had dropped by 60 percent.

After years of uncertainty, oceanographers made the link to more corrosive acidic waters at higher levels in the water column. Acidification robs young oysters of the minerals they need to make their shells. Without abundant materials, the young oysters work too hard, exhausting themselves and making them prone to disease and die-off.

Hatcheries have invested in high-tech pH sensors to track and cut down on the acidity of the ocean water they pump into their operations. This added awareness helped them quickly change practices, which led to an oyster rebound.

However, the ordeal sowed doubts about the long-term viability of the industry. Taylor Shellfish responded to the threat, embracing technology to provide clarity on changing conditions and add operational resiliency.

The company's digital transformation is unfolding against a backdrop of growing demand and reduced output from wild fisheries. The transformation puts aquaculture at the forefront of maintaining sustainable seafood production to feed a growing population.

Tracing oysters

Taylor Shellfish Farms, a fifth-generation family-owned aquaculture operation based in Shelton, Washington, has steadily expanded the scale and scope of its operations over the years. It started in 1890 when the family began farming shellfish in the Puget Sound, and one farm sold the catch to a processor. In the next generation, the family expanded to processing shellfish. More recently, it has added hatcheries and its own oyster bars to sell products direct to customers.

The journey of each shellfish starts in the hatcheries where Taylor Shellfish Farms breeds oysters and clams, including geoduck. These "seeds" are then planted on many tidal beaches. The company stewards the shellfish to maturity and then harvests and processes the

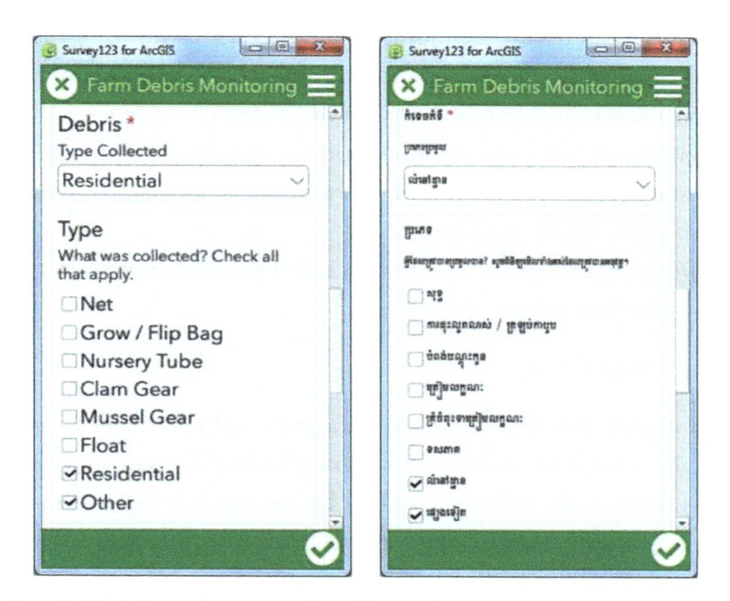

Taylor Shellfish Farms has created multilingual surveys in English, Spanish, and Khmer, the Cambodian language (right), to help make data collection a habit for diverse crews.

shellfish. Finally, the company distributes the shellfish, bringing the product to market and to the tables in its oyster bars.

Apps to keep track of each farm bed's contents and maturity are central to the company's genetics program. Taylor aims to track the farm's processes, with the benefit of being able to correlate the relevant practices in each oyster's upbringing. This knowledge will help Taylor with the nature or nurture questions that arise in genetic selection.

"We have shuckers that open hundreds of oysters every day in our oyster bars, and they can tell us which ones aren't looking good," said Nyle Taylor, farm project coordinator and fifth-generation family member.

This approach combines human sensors, who are experts in shellfish quality, with technology that aggregates information on what happened where. GIS provides this backbone of capability, and farmers use handheld apps to inform the system.

With knowledge of conditions and lineage, Taylor staff can pick the best oyster offspring, looking for resilience to acidity, the greatest growth rates, appearance, and taste.

"We grow enough oysters that a 2 or 3 percent improvement in survival has real value," Taylor said.

Sharing knowledge

The company provides apps that its farmers and fieldworkers use to track operations on all the farms, from planting the seeds and moving the crops to the maintenance that happens as harvest approaches. These apps communicate with a cloud-based GIS, which gives workers the ability to visualize and analyze the data and detect operational inefficiencies.

"We can compare farm to farm, understand the techniques that lead to improvements, quantify that value, and push the best techniques to other farms," Taylor said.

Apps also help keep track of inventories, budgets, and various operational details. The company must comply with regulations at the local, state, and federal levels that require many permits, and most permits require a map.

For years, the company would submit hand-drawn maps for every permit, but this practice changed when Taylor hired Erin Ewald as its assistant director of regulatory and environmental compliance. Ewald had experience with GIS and used it to maintain maps of farm beds. She updated the maps with GIS and made the data mobile.

"Now, we can compile information and push it to the right people," Taylor said.

Easy-to-use apps make this transformation possible. "These apps are so straightforward that our farmers see the benefit," Taylor explained. "They can use them while they do the work, saving them time, and they don't have to come to the office to enter data when they're all wet and muddy from the tide."

Stewards of land and people

The company adopted environmental processes to ensure farming practices work well with the natural environment in Puget Sound, careful not to harm the salmon and forage fish species around their farms.

The organization prides itself on its long history of environmentalism, achieving Aquaculture Stewardship Council certification in the United States that designates the shellfish as responsibly farmed.

Taylor's leadership also thinks about its people and its land for the long term.

"Sustainability isn't just about the environment for us," Taylor said. "It's also about operating responsibly with our employees and in our communities. Paying living wages and providing benefits are important to us, as is making sure our workplace is safe."

Recently, one of their farm directors was in a serious car accident and suffered memory loss as a result of his injuries. But mobile apps that capture farm details and processes provided the farm manager with a backup, putting the information about the farm in his hands and assuring him that he can pull up exact details when needed.

"We've reached a size where having the information in everybody's heads is not the best way of doing things," said Taylor. "Our digital transformation has allowed us to bring each farmer's knowledge into a shareable system that can be passed from generation to generation."

GIS and handheld apps power digital transformation

Taylor Shellfish Farms uses apps to take real-time data on business, operations, and environment to its 30 farms. The ability to view GIS data in the field helps the company understand the complexity of its operations.

The apps allow farm managers, with years of experience, to share real-time changes in conditions and record suggestions for where each farm could expand. The offline editing capability in ArcGIS Explorer serves the company well, partly because many of the farms are in rural areas with spotty cell signal coverage. Explorer provides a handy repository for data and attachments, such as permits that can be shown to fish and wildlife inspectors when needed.

Taylor uses Survey123 for field data collection, such as reports of shoreline debris or the presence of herring spawn, which it's required to report to regulators. It has created multilingual surveys in English, Spanish, and Khmer (the Cambodian language) to help make data collection a habit for diverse crews.

The company uses ArcGIS Collector to update operational data about its farm beds. The data is used to evaluate the status of various farms, freeing up information that was previously only in the minds of farmers.

ArcGIS Workforce helps the company collaborate with the local Squaxin Island Tribe in Oakland Bay, Washington, which has been harvesting shellfish there for centuries. For example, more than an inch of precipitation requires water sampling. When that happens, the company can select and assign sampling points—for instance, by sampling marine water and assigning freshwater sampling to its tribal colleagues.

ArcGIS Drone2Map® has improved the mapping of farm beds. Drones can capture images and determine the elevation of beds at low tide, so that farmers know when the beds are at optimum depth for farming each species. The company can better see beach drainage and lay out the beds so the seed won't get washed away. Finally, 3D maps help Taylor Shellfish Farms share its sustainable "Tide to Table" story with its customers.

A version of this story by Caitlyn Raines originally appeared on the *Esri Blog* on September 21, 2017.

PART 2

MODERNIZING FORESTRY

FORESTS ARE COMPLEX, DYNAMIC LIVING SYSTEMS. BY understanding forest science, economics, environmental considerations, and societal desires for cleaner and renewable resources, organizations can better care for the forest and meet business goals more sustainably. Embracing technological advances and modernizing forest management practices help foresters transform how they collect, analyze, and use authoritative data to drive improved decision-making. Today, more forestry organizations than ever are using GIS to deliver real-time operational intelligence and improve forest health.

Forestry organizations of all sizes and types are embracing digital innovation to be more successful and sustainable as follows:

- **Public forest agencies:** Managing public forest lands is becoming increasingly difficult, considering changes in climate, population, environmental and ecological pressures, and public expectations. Agencies must manage, preserve, and restore forests, which requires a careful balance as they strive to create a sustainable future.

- **Commercial forestry organizations:** Private forestry companies must balance economic profitability with their responsibility to staff, the communities they serve, and future generations. Digital innovation helps support this balance by improving operational efficiency and decision-making.

- **Partners and community outreach:** The diverse forestry community has many workflows, concerns, and interests. Technology partners who have built solutions that service these needs play an integral role within the community to support industry workflow challenges.

Next, we'll take a more detailed look at how forestry organizations can use GIS to support data collection, operations, sustainable business practices, forest interfaces, and planning for the future.

Planning and data collection

GIS helps modern forestry organizations plan and manage their field operations more efficiently. Collecting field and remotely sensed data and analyzing it in near real time using performance dashboards helps foresters visualize complex workflows, accelerate response times, and reduce costs. Organizations can use location technology to manage forest inventory and plan resources more precisely while improving quality and customer satisfaction. Automating the processing and analysis of collected data also helps organizations monitor tree growth and forest health.

Managed operations

Today, foresters are replacing outdated workflows to measure stands of forest, conduct silviculture assessments, and inspect timber assets with efficient and easy-to-use applications for collecting and recording field measurements and observations. They are integrating these spatial datasets with business data to better understand current forest conditions and trends. The use of spatially sensitive analytics brings business workflows to life and drives better decision-making.

To succeed, organizations must reduce operational costs while increasing efficiency through greater situational awareness, workforce coordination, real-time mapping, and spatial analysis. Achieving these goals allows forest managers to meet planned objectives, compliance objectives, and budgets. Their ability to quickly visualize information can prevent or mitigate damage to harvesting operations and forest lands and deliver information to decision-makers and stakeholders to help them make decisions.

Sustainable business practices

Many forests today are managed to provide balanced multiuse ecosystems, including recreation and tourism over prolonged periods. Various types of forests are a valuable resource in scenic landscapes that support diverse natural wildlife habitats (including migration corridors, refuges, and fish habitat) that can provide licensed hunting, fishing, and organic food gathering. The presence of forests can help mitigate the growing pressures on wilderness areas and the wild and scenic rivers they contain. Mitigation requires a holistic approach to managed forest health, including understanding and compromise under a shared stewardship model. The best of these new models show the value that forests provide to regional watershed services: clean water, erosion control, surface-flow regulation, and stream bank stabilization for present and future generations.

GIS can help multiple stakeholders collaborate toward balanced multiuse forest management practices for improved outcomes with broad-scale importance to long-term sustainability. GIS can also help people understand the balance of careful and selective forest harvesting in a wider context, considering the holistic value of healthy forest ecosystems, and determine more equitable solutions for long-term forest sustainability under shared stewardship models—for example, by tribal, environmental, and recreational organizations.

Managing forest interfaces

As forest boundaries face pressures from economic development, the value of these resources can be readily eroded. We need models that respect a more holistic approach to forest management that addresses these questions:

- How should future fire protection plans incorporate the urban interface and forest communities in a potentially drier climate?
- What are our expectations for wildland fire management in the future?
- What investments are we prepared to make to better manage fire-adapted landscapes?
- How should tribal concerns and important cultural landscapes be considered?

GIS can facilitate planning to understand, analyze, and report on forest urban interfaces and the impact of proposed development and change in land use. Economic development and the protection of natural habitats and ecosystems require an equitable balance. GIS supports wildland fire management strategies, including preparedness, suppression, and rehabilitation. Specifically, GIS provides firefighters with tools to identify, analyze, and understand the landscape; maps provide the situational awareness firefighters need to save lives and protect property.

Planning for the future

As the world moves to a low-carbon strategy, what part will forests play in terms of carbon sequestration? Preliminary models are being devised for cap-and-trade markets. Organizations are looking for solutions to reduce their own operational carbon footprint by using interactive reports and dashboards to inform and involve stakeholders and support compliance.

GIS can help organizations consider optimal areas for carbon sequestration projects and determine how they overlap with managed forest lands. GIS supports remote sensing and field data collection to evaluate, promote, and share prospective project sites and model spatial and temporal changes to forest carbon stock. The interactive tools of GIS can help organizations and stakeholders analyze project options and risks in the context of increasing demand, climate change, and forest ecosystem insecurity.

The next section presents several real-life stories about how forestry organizations are using GIS to modernize forest mapping to create more efficient operations with a holistic, long-term view of forest management and sustainability.

LUMBER MILL TAKES GUESSWORK OUT OF OPERATIONS USING GIS

Culp Lumber

THE SOUTHERN YELLOW PINE FORESTS ACROSS THE southeastern United States provide the country's primary source of construction lumber. The wood is strong and dense, and it holds nails well. It was also plentiful and cheap until the recent pandemic-related home remodeling trend. In response to increased demands, lumber companies are running mills at capacity, while reveling in a 280 percent price spike after many lean years.

That demand hasn't dulled efforts by one lumber company to keep designing better ways of doing business. The company has remained committed to sustainability by using high-tech methods in a decidedly analog place—the forest. In an industry accustomed to slim profit margins, H. W. Culp Lumber Company, based in New London, North Carolina, has been using location intelligence to make more informed decisions about where it harvests and how far it will drive to get the lumber it needs as fuel costs rise.

"When you're buying [and transporting] a half million tons a year, that becomes real money really quick," said Chris Charest, procurement forester at Culp Lumber.

The latest technology adoption comes years after the third-generation family-owned business, in operation since the early 1920s, first automated its mill operations in the mid-1980s. And it continued to focus on increased efficiencies and new technologies to survive the Great Recession that began in late 2007.

When logs get to the mill, machines take over—independently scanning, assessing, and sorting to maximize profits.

"Once it's in the mill—unless there's something that goes totally haywire—a human doesn't make a decision," said Charest. "The computer will cut a 16-foot board in half if it will make a penny or more difference in price." The resultant products include yellow pine lumber, timbers, boards, and by-products, including sawdust, shavings, and chips.

Every part of the tree is valuable, but there's a greater value for a No. 1 board with its dense grain pattern and smaller, tighter knots. A computer at the mill does the grading, using machine vision to assess and sort boards at a pace of 150 pieces per minute.

Charest marvels at the efficiency gains in the mill and sweats the tight margins in his role as procurement forester. Now, he's working to tie a modern GIS to location-aware apps to streamline workflows in the forest—from assessing to cutting to hauling. The field apps collect data, feeding the analysis of forests to determine whether a specific timber stand can be logged profitably. A lot comes down to the miles of travel.

Feeding the mill

The Culp Lumber mill operates four and a half days a week with continuous sawing for 10 hours per day. Before the Great Recession, the mill produced 425,000 board feet per day, and now it produces 700,000. To keep up with the pace of production, 100 tractor-trailer loads of pine sawlogs are required every day.

This part of the country has plenty of supply. Even though Culp Lumber harvests trees only within a 100-mile radius of its mill, the pace of pine growth and amount of pine forest in the area can more than keep up.

"We're seeing the benefits of the Conservation Reserve Program when the federal government paid landowners to take land out of agriculture production and put it in trees," Charest said. "And the trees are growing faster than we're cutting them right now."

The Culp family owns several thousand acres of forest that provides some of the wood it mills, but the bulk of it is bought from

Culp Lumber Mill in New London, North Carolina.

other landowners in a process that involves sealed bidding. The landowner sets the sale date and provides a map of the tract that the bidders appraise. On the date and time of the sale, the bids are opened and the highest one gets to log the tract.

Charest manages a group of five foresters who look at the trees in a process called timber cruising to find the right logs within the shortest distance.

"We generate cruise points with systematic spacing for every forest we evaluate, and we can manipulate the layout on the computer through GIS before we leave the office, so we know what we're going to do before we hit the woods," Charest said.

Because Culp Lumber specializes in yellow pine, Charest looks to buy tracts with 60 percent or greater of the species. In the forest, each worker measures trees at the center of the plot, recording the diameter, height, species, and stem quality. The statistical measurements of trees are then fed into a program in the office to generate lumber volumes that Charest uses to determine the value for the stand of timber.

The control room in Culp Lumber Mill.

Awareness to improve trucking costs

One of the next workflows Charest is addressing is distributing maps to the loggers so they can see boundary lines and know where they are on a tract of land. The aim is to ensure that all (and only) trees bought get harvested. "Cutting over on someone else's timber gets expensive," Charest noted.

Most logging crews are down to a three- or four-person operation with heavy equipment. The feller buncher is the cutting machine with a big saw head that can drive through a tree, pick it up, and lay it down in bunches. Then the skidder picks up the bunch and drags it to where the loader will delimb it and throw it on a truck.

Culp Lumber knows when a truck has been loaded and how many logs are coming in. But routing the trucks and paying the drivers for the miles they've driven required the more detailed maps produced using GIS. A recent data upgrade with the location of low-weight bridges made a big difference in trucking costs.

"If we pay 15 cents a ton-mile," Charest continued, "let's say we buy a tract of timber, and it's 40 miles back to the mill. That's 40 times 15 cents, for $6 a ton. If there's a low-weight bridge, we're legally not allowed to cross it. So, if it's 10 miles to go around it, all of a sudden, I have to add $1.50 per ton to the delivery price. If the tract is 10,000 tons, then it's going to cost us $15,000 more because of one low-weight bridge."

Charest always has a calculator close at hand because the delivery price is crucial to answering his key procurement question, What am I willing to pay? Calculating trucking costs correctly before the timber is bought allows him to determine what he can pay the landowner and still make a profit.

"The ability to build a special routing program based on the parameters that we need for our log trucks became huge for us," Charest said. "As fuel rates go up and the truck driver shortage grows, miles are going to become more important to companies."

Forest and the family

Culp Lumber participated in the Sustainable Forestry Initiative (SFI) for 15 years and credits the process with instilling good procurement practices that the company continues to follow.

"The intent is to keep it a family-run mill, so we are concerned about the longevity and availability of the product in our wood basket," Charest said. The wood basket refers to Culp's harvest area as well as the southern states that have been producing lumber for centuries. The wood basket region stretches from eastern Texas to Maryland and produces 60 percent of the country's wood products.

The US timber industry lost more than 70,000 jobs after the Great Recession. Mills closed, and those that were still running operated at limited capacity. At Culp Lumber, the family-oriented focus helped get it through the tough times with everyone pitching in to find efficiencies and scrape out a profit to keep everyone employed.

"This is a real family-oriented culture," Charest said. "We have 40-year employees, 30-year employees, 20-year employees. We have employees that started working here before they got out of high school, and they don't go anywhere until they retire."

The culture provided motivation and kept everyone asking how to improve throughput and profit in the mill. The latest wave of change in process at Culp Lumber involves digitalization to bring information together to tell a complete story and involve the entire staff.

"A lot of information is kept in my head and my boss's head, and the owner recently asked what would happen if [we] 'met the beer truck' one day," Charest said, referring to their untimely demise. "The answer was, 'That would be a very bad day for Culp Lumber.' So, now we're looking at how to bring everything together so everyone in the organization understands what's going on."

Charest has embraced modern GIS and the location intelligence it delivers, providing enhanced employee awareness through a shared map. He hopes to bring efficiency gains to the procurement side of the business such as those the company has seen in the mill.

"I'm starting to track bids we lost and going back to run the analysis on why we lost them," Charest said. "Before, we just threw them in a drawer. We're making the information more readily available and usable so we can look for any little advantage, because at the volumes we're running, every little thing you do can translate to huge numbers."

A version of this story by Audrey Lamb titled "North Carolina Lumber Mill Takes Guesswork Out of Operations with GIS" originally appeared on the *Esri Blog* on July 13, 2021.

HUB BUILDS COLLABORATION FOR FOREST PLAN

Montana Department of Natural Resources and Conservation

MONTANA'S APPROACH TO IMPROVING FOREST HEALTH and reducing wildfire risk statewide includes a web-based location platform in which the state brings together forest data and makes it accessible. Forest stakeholders used the platform to collaborate on completing the 2020 Montana Forest Action Plan.

Good stewardship and management of forests is essential to many Montanans: recreationalists; the forest products industry; federal, state, tribal, and local-level land managers; private forest landowners; representatives of conservation organizations; collaborative and watershed groups; ranchers and farmers; wildlife watchers; and other partners. All these have vested interests in the health of the state's forests. Although these stakeholders value the state's 23 million acres of forested land in different ways, they all have a common goal: keep Montana's forests healthy and resilient.

Under the authority of the 2008 and 2014 Farm Bills, Congress tasked states and territories with assessing the condition of the forests within their boundaries, regardless of ownership, and developing strategies to promote forest health and resiliency through a state forest action plan. Montana's first plans were static hard-copy or digital documents that left little room for change or making iterative versions.

When it came time to revise the Montana Forest Action Plan in 2020, the Montana Department of Natural Resources and Conservation (DNRC) became the principal agency responsible for the revision of the state's forest action plan. DNRC wanted to use technology that would ensure the plan's continued relevancy.

Montana Department of Natural Resources and Conservation revised its state's forest action plan using ArcGIS Hub to ensure the web-enabled plan is easy to access, explore, and update.

Montana's web-enabled plan would be a living document—easy to access, explore, and update. By using ArcGIS Hub℠ applications, the 2020 revision incorporates the most up-to-date data and science as it becomes available, thereby providing accurate and timely information as Montana changes over the next decade.

More importantly, Hub will show, at landscape scales, how the Montana Forest Action Plan has changed forest health and wildfire risk and communicate that information in ways that are accessible and easily understood.

DNRC's GIS team had experience developing web mapping apps, which bolstered revision efforts. They had previously created the Montana Interactive Wildland Fire Information Tool for sharing wildfire information with fire managers, decision-makers, and the public. This web app proved its worth, particularly during the intense 2017 fire season, when thousands of individuals used it daily to see up-to-date information about fire conditions.

DNRC GIS manager Brian Collins thought a similar solution would work for the 2020 Montana Forest Action Plan. This time, he wanted to use GIS to build an information center for participants to share their data, ideas, and goals.

Collins knew that buy-in from top department executives, who make decisions and build policies, would be critical to the project's success. He set up a meeting with Montana's State Forester Sonya Germann, administrator of the DNRC Forestry Division. Collins pitched the idea of implementing GIS to serve as a focal point, which would encourage collaboration and engagement with the Montana Forest Action Plan.

"The 2020 Montana Forest Action Plan is an opportunity to use geospatial technologies that improve planning, reporting, and understanding," Collins said.

Collins described GIS capabilities to Germann by demonstrating the functionality of ArcGIS Dashboards and ArcGIS StoryMaps℠. He then introduced the concept of a cloud-based engagement platform with a hub for sharing information between departments and engaging the public. It could be the venue for planning and collaboration.

Hub manages content and data and can display them as maps, dashboards, ArcGIS StoryMaps stories, documents, and website pages. Organizations use hubs to gather data for their projects and contribute their own data for others to use.

During the Montana Forest Action Plan revision process, staff and relevant partners used Hub to collaborate and share information and ideas. The public and important forest stakeholders also used Hub to engage in conversation and submit feedback.

"I explained how ArcGIS Hub consumes different types of data and then represents that information through different means, so it connects with people in ways that make sense to them," said Collins. "When it comes to communicating data, we need to make it as accessible as possible, and ArcGIS Hub allows us to do that."

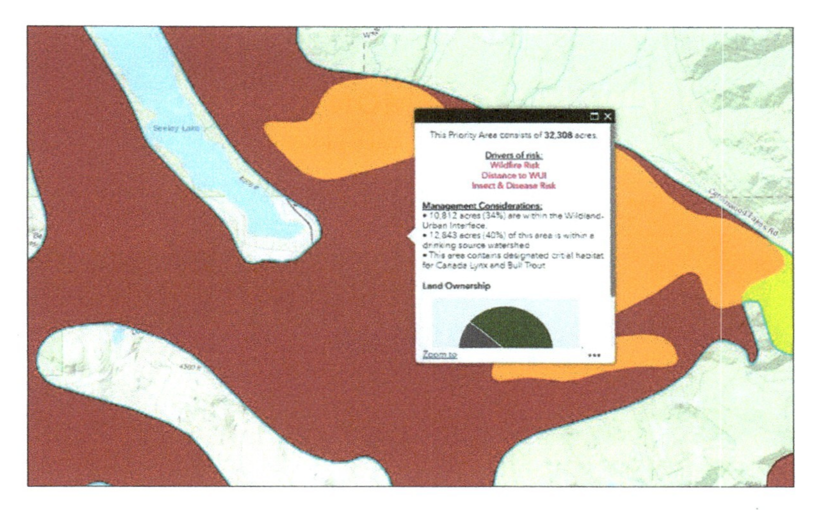

One of the dashboards showcased in the web-enabled plan that identifies priority areas with dynamic pop-up content, such as risks and management considerations.

State Forester Germann saw value in the technology and immediately got behind the project. The GIS platform and Hub would support the Montana Forest Action Advisory Council. The council is a group assembled by Montana's governor to help develop the Forest Action Plan and implement strategies to help improve and sustain forests throughout the state. Germann also made certain that the GIS manager played a key role on the project's core team.

The DNRC GIS team implemented Hub and used the department's existing open data to make and share maps. Data resources grew because agency partners and council members began to share authoritative open data via Hub.

Lead geospatial analyst for the Forest Action Plan team Nick Youngstrom said, "Hub has enabled us to reduce the friction between data and the people that need to be informed by it. We've removed some of the technical knowledge and staff power needed to use and access geospatial information on Montana's forests."

The Montana Forest Action Plan uses ArcGIS StoryMaps to explain the concept of cross-boundary forest restoration management.

All the council's data, as well as interactive maps and additional information, is available on the Montana Forest Action Plan website (montanaforestactionplan.org). Visitors can explore data layers used in the plan, such as wildfire hazard potential and an interactive analysis of Montana's urban forests. Participants can also dig deeper into the plan's data, using the Hub dashboard to visualize and understand information and track progress toward the accomplishment of goals and objectives set by the council. The GIS team created mapping apps to help the public better understand where projects are located and the specific goals or purposes of those projects.

Additionally, the website hosts an ArcGIS StoryMaps story that presents the plan's initiatives. For instance, to help illustrate the Shared Stewardship initiative, the "Finding Common Ground" story explains the concept of cross-boundary forest restoration management. Although the topic sounds daunting, the story format makes the concept understandable and engaging.

"Hub technology made this planning process, by far, the most successful effort we've had for collaborating on data collection efforts with a multitude of partners," Collins said.

The Montana Forest Action Plan, completed by fall 2020, is a living document that can evolve with the times. Its platform serves as a nexus of information, analysis, and engagement and ensures that all Montanans have the information and resources they need to keep Montana's forests healthy and resilient.

A version of this story originally appeared on esri.com in 2020.

DOCUMENTING WORKFLOWS FOR SUSTAINABLE FORESTRY

Starker Forests

IN 2020, WHEN THE FOREST-RICH STATE OF OREGON amended the Oregon Forest Practices Act under Senate Bill 1602 to regulate how and where herbicide and pesticide may be sprayed on private forestlands, it required forest landowners to share activities in near real time.

At Starker Forests, a fifth-generation family-owned land company located in Corvallis, the regulation came as the business was implementing information technology updates. Foresters at the company have used GIS technology for decades to map their work, but it had been primarily an individual tool without much collaboration. With the reporting requirement, Starker decided to adopt Web GIS, with all foresters sharing data about the forest and their work.

"There are a ton of possible systems, but so much of what we do is tied back to on-the-ground geography that, no matter what we use to collect, store, and manage our activities data, we were still going to be using GIS," said Rick Allen, reforestation forester at Starker.

Centralizing the mapping

Allen is one of several foresters at Starker Forests who are soon to retire, prompting the company to make improvements to its data collection systems. The idea had been to streamline the company's recording system to capture the institutional knowledge of this cadre of professionals—who, during their careers, had achieved sustainability certification from the American Tree Farm System—before they left the company.

Oregon's new rules reinforced the need to evolve from a paper-

Wood products keep carbon in storage and offer a more sustainable alternative than materials such as concrete and steel for residential and commercial construction.

based process, which the company had used for decades, to a faster mode of record keeping. "We wanted to automate the process and eliminate how paper tends to stack up," Allen said.

Allen made the shift to take on the IT challenge after spending a career in reforestation, which includes managing forest site preparation, tree planting, and herbicide applications. Herbicides are used in the first few years of a forest plantation to hold back invasive species and other vegetation that competes for sunlight and moisture with planted seedlings. "The company prioritizes the reforestation process as an investment for the future. Getting new trees in the ground within a year after harvest is the goal so that they're off to the races," Allen said.

With the 2020 legislation, the Oregon Department of Forestry extended buffer zones to keep the spray further away from homes, schools, and streams. Forestry is one of the most heavily regulated industries in the state. The extended buffer zones further protect people, drinking water, and fish habitat.

Allen sees the value of better spraying documentation for Starker's own purposes. This new system will allow any of the professional foresters to quickly locate and check the historical records for the herbicide applications.

"We found that trying to go back in time over paper records to figure out what nursery stock was planted and from which nursery, weather conditions, what chemicals were applied, and the application rates could be tedious and hard to figure out," Allen said.

Advancing enterprise record keeping

Starker had investigated investing in an enterprise resource planning system—a large database—but opted to build new tools using GIS.

With the commitment to GIS, the Starker team worked alongside Esri® Professional Services to build field apps for specific workflows. The mobile-first approach meant that the tools would be used first in the forest to capture details specific to the different operational phases, from harvesting to reforestation.

Addressing end-to-end workflows with tailored tools often starts with Survey123, an app with simple forms that record data that can then be analyzed in the office. The data also feeds dashboards built with Dashboards to show progress toward goals, such as the yearly planting of more than 500,000 trees.

Starker foresters manage more than 90,000 acres of coast range forests, including Douglas fir, Oregon white oak, western red cedar, western hemlock, ponderosa pine, Oregon ash, and grand fir. The company doesn't own a mill or the mechanized equipment used to harvest trees. Instead, it uses contractors to harvest trees on approximately 1,000 acres per year, with some thinning and clear-cutting.

With the new tools, the foresters have saved time on repeatable workflows and gained a better understanding of the health of the forest. Every year, hiring ramps up in the summer for forestry interns

Starker Forests follows the best management practices outlined by the American Tree Farm System. Signage points out management practices, such as thinning, that lead to a healthy forest.

who take an inventory of portions of the forest. They check for survival of the trees planted in prior years and gather details on competing vegetation that might trigger the need to apply herbicide to help the seedlings survive.

"A dashboard helps us quickly look at thousands of plots to see where we have issues," Allen said. "We share that information with our board and use it to make management decisions."

Managing sustainable harvesting practices

Starker Forests is certified by the American Tree Farm System, which audits and validates the company's management practices. The certification process ensures that the company's management plan improves air, water, and soil quality while managing habitat for wildlife and cutting back on invasive species.

Water quality is one of many environmental considerations that Starker carefully manages. The cable logging system the company uses pulls cut trees to the top of ridges and away from streams and wetlands. Road construction and maintenance are improved to minimize erosion to keep water clean. Buffers are maintained for spraying and harvesting to protect streams.

Allen and the IT team are working now on harvest planning and scheduling workflows to improve those processes.

"Everything starts with the harvest, and then all subsequent workflows follow—site preparation, spraying, tree planting, and

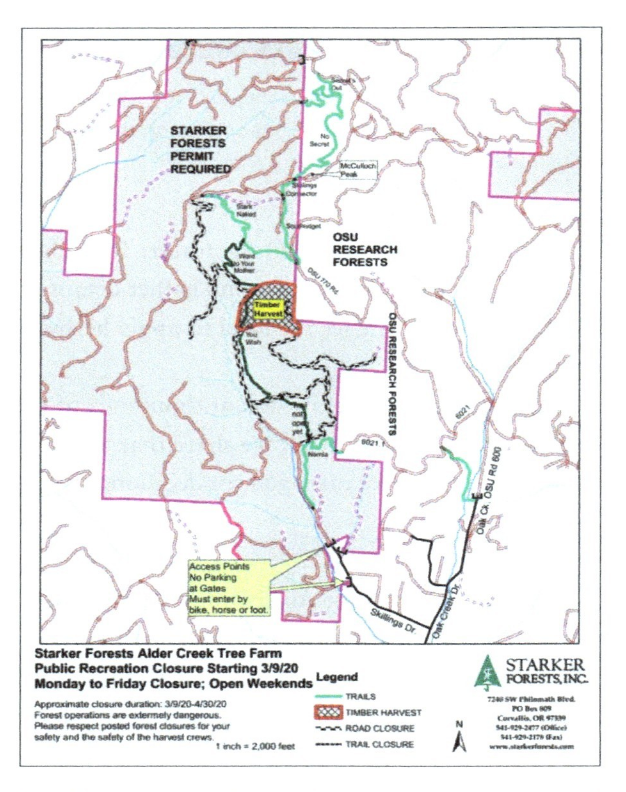

Map marks public recreation closures during harvest operations at Starker Forests Alder Creek Tree Farm.

thinning," Allen said. "With the dashboard, the person responsible for each step can look and see when each process will be done for each unit so they can make sure a contractor is scheduled to move in when the prior work is finished."

Starker has a long history of sustainable forest management and is focused on improving tracking of sustainability metrics, with a dashboard that tracks seedling survival and growth, silvicultural treatments for young stands, and wildlife projects. The dashboard and data it collects demonstrate sustainable growth and harvest levels across the acres Starker owns. This allows the company to identify areas for improvement and plan for future generations.

Under a stewardship agreement with the State of Oregon, a landowner that shows commitment to protecting natural resources, such as water and fish and wildlife habitat, can gain long-term regulatory certainty without fear of further restrictions by sharing evidence that it follows Forest Practices Act rules. Beyond better planning, the enterprise-level GIS that Starker has deployed makes it easier to monitor and report the actions the company takes.

"It has been a process of awakening," Allen said. "We identified early on that what we needed was more integration, moving from a culture of independent work by 12 people with 12 different workflows to a single repository where we can see the status of the forest and all our work."

A version of this story by Scott Noulis titled "Starker Forests Documents Workflows for Sustainable Forestry" originally appeared on the *Esri Blog* on August 23, 2022. Images courtesy of Starker Forests.

ARTIFICIAL INTELLIGENCE AND ROBOTS ENHANCE FOREST SUSTAINABILITY

Finnish Forest Centre

FINLAND'S FORESTS COVER THREE-QUARTERS OF THE country's landscape—a vast space that the Finnish Forest Centre is responsible for monitoring and managing. Recently, the center's forestry experts developed an ambitious idea to use robots to autonomously perform most forest maintenance tasks.

Before they can train the robots, Finnish Forest Centre staff must be sure their data is accurate and accessible. They use aerial and lidar imagery to create forest inventories, cataloging tree measurements and species details. Then they collect, map, and analyze all the data that robots will need in a GIS.

Modern forest harvesting machines are increasingly capable. Here an operator uses a PONSSE Ergo forest harvester on a winter morning in the south of Finland.

"There's no reason why tree harvesting machines can't be automated," said Tapani Hämäläinen, development director of the Finnish Forest Centre. Hämäläinen and his team see robots as providing several benefits. Automation could simplify forest management, especially for private owners, who hold more than 65 percent of Finland's forests. It would also allow proactive management of forests, a move that would maximize the carbon sequestration of trees. And robots could fill an employment gap due to a current lack of interest in forestry jobs.

"Going to the forest for 12 hours a day to drive a harvesting machine is lonely work in a dark forest," Hämäläinen said. "Some people like that, but there's no queue of people who want to be trained for these jobs. It's a real problem that can't be solved without automation."

Freedom and Finland's forests

In Finland, a private landowner may not restrict people from hiking through the forest or harvesting berries, mushrooms, or wild herbs there. *Jokamiehen oikeudet*, roughly translated as "freedom to roam," gives anyone in Finland access to the forest.

In addition, Finland's constitution gives private property holders the right to do as they wish with their land.

"That's why, when you want to cut your forest, you don't ask for permission, you simply make a declaration," Hämäläinen said, "although they do have to give us the information."

Finland's private forests

Weather and a long history of sustainable forestry make Finland the most forested nation in Europe. Revenues from the forest industry have contributed nearly 30 percent of Finnish export income, much of it from pulp and paper production.

In 1994, Finland adopted new environmental regulations, balancing the need for a sustainable economic return with environmental preservation and biodiversity. Foresters abandoned practices such as draining bogs or applying herbicide to kill undergrowth and started planting four trees for every one cut down. This same legislation also encourages landowners to take an active role in managing their forestland.

"We want to activate forest owners," Hämäläinen said. "They may have inherited a piece of forest and don't know where or what they hold, so they don't do anything."

The Finnish Forestry Centre provides private forest owners with forest management plans and recommendations. Center staff help owners calculate the value of their forests to transact a sale or collect payments for conservation. The center also provides subsidies for planting seedlings or building infrastructure such as forest roads or drainage ditches. In a country as lush as Finland, this level of attention to forests is a must.

Selective harvesting gives remaining trees room to grow.

"Doing nothing leads to a forest so thick that it doesn't grow or take in more carbon," Hämäläinen said.

Automation starts with accurate data

The Finnish Forest Centre has created a website, Metsään.fi (meaning "the forest"), that delivers e-services and creates a marketplace for forest owners to connect with forestry professionals for services including logging work. The site provides open access to (and download of) the government's forest inventory data with transparent measurements that both the buyer and seller can use.

"We collect all the forest data nationwide," said Juha Inkilä, a data specialist at the Finnish Forest Centre. "We conduct a continuous inventory that cycles through the entire forest every six years."

Center staff update the portal with data obtained from laser scanning, aerial photography, sample plot measurements, and site visits. Between surveys, information such as work reports is maintained based on notifications received by the Forest Centre from forest owners and forestry organizations. Tree growth is factored into portal data, with suggested actions updated annually.

Now, the Finnish Forest Centre serves as a central source for accurate, authoritative, real-time data used by forest owners and others in the industry.

"We want the data to be so accurate that users can see and sell without having to go to the forest," Hämäläinen said. "In 85 percent of the cases, we have accurate data now, and we're working to solve the final 15 percent."

Machine learning to aid forest understanding

To fill the gap on accurate data and decrease the amount of work on-site, the Finnish Forest Centre is using AI. One of its first use cases for AI is to gather information about young forests and seedling

Much of the harvest goes to pulp mills to make paper products.

stands. GIS data, imagery sources, and climate and weather datasets are combined to produce accurate measurements of forest stands and better predict forest inventory. Machine learning algorithms are being tested to extract the same measurements that have previously required in-person inspection crews.

For foresters and landowners, the work of AI will inform them about the volumes and species of wood in a forest without the need for a costly, time-consuming inspection.

AI and big data will also be used to detect signs of spruce bark beetle attacks, detected on satellite photos of forests, and to optimize logistics and the supply chain.

In the 2019 Finnish elections, the Green Party became the country's second-biggest political party, raising a new level of interest in the forests.

"Sustainability is a big thing globally and also in Finland," Hämäläinen said. "Of course, the government wants to change legislation toward preserving more carbon, but they don't yet have specific plans about how to do that and what would be the best way to do it."

The issue holding back a change in forestry legislation is that science has yet to determine the best mix of practices to maximize the carbon sequestration capacity of trees.

"The biggest questions right now are, How much carbon does the forest take? And is it better to harvest selectively or clear-cut and then plant new trees?" Hämäläinen said. "At the moment, you get different answers, but eventually our data will reveal the correct answer."

A version of this story by Scot McQueen titled "How Artificial Intelligence, Robots Enhance Forest Sustainability in Finland" originally appeared on the *Esri Blog* on December 16, 2019.

PART 3

SMARTER MINING

LOCATION INTELLIGENCE IS FUNDAMENTAL TO THE mining life cycle, from mineral and metal exploration to mine remediation. Efficient and profitable mining companies need detailed information to meet the global demand for resources on which a growing world economy depends. Because minerals and metals are depleted faster than they can be replenished in many areas, the strategic and environmentally sustainable development of new reserves requires an understanding of spatial context.

With tools to compile, process, display, analyze, and archive massive volumes of data, GIS technology is increasingly being applied to the business of mining for these purposes:

- Mining professionals use GIS software to increase productivity, reduce business risk, and save costs.
- Engineers use GIS for mine-planning applications that help monitor existing infrastructure, integrate up-to-date information with the mine plan, and keep workers safe.
- Facility managers incorporate survey data, IoT data from environmental sensors, and imagery from drones to provide better contextual oversight of their broad and complex operations.

Most mining information, including financial and asset information, has a spatial component that, when represented in map form, provides greater management or operational context. Managers and mineral economists use GIS to evaluate corporate and competitor assets, consolidate information, and choose accurate business strategies.

GIS provides direct access to data in corporate spreadsheets and relational databases. Reserve estimates, annual planned production, and cost-per-ton statistics can be linked to prospective sites or existing mine locations. Placing these sites in a geological, political, and economic setting can support regional mining exploration. Detailed exploration prospects and active mine data can be accessed through intuitive interfaces and easy-to-navigate visual displays.

Many mining companies draw from the vast amount of GIS data on the internet. Companies can distribute map and image data throughout the enterprise or on the web. This capability allows mining professionals to share information in real time across global corporations.

Let's take a more detailed look at how mining organizations can use GIS to support the most common application types, in exploration, operations, environmental health and safety, and logistics workflows.

Exploration

Successful programs for metal, mineral, and aggregate exploration require seamless access to data, stakeholder collaboration, and intelligent processes so leaders can make timely and informed decisions and manage business risk. GIS enables staff to share information with internal and external stakeholders, combine disparate datasets, integrate workflows, and gain insights into exploration prospects.

GIS allows mining operations to explore, manage, and aggregate data in a common and integrated platform. For example, you can build your strategic land-use plan to conform with regional environmental legislation and identify geologic variables that will guide project viability.

An effective and optimized field program can be constructed using strategic elements from initial prospecting observations. You can use GIS tools to collect data rapidly and seamlessly in the field and merge it. The ability to merge and integrate data into a common platform is paramount to successful exploration.

To synthesize field data, geoscientists use mapping and analysis tools to visualize, share, and communicate field observations in near real time. Using GIS from a common enterprise platform and collaborating with partner technologies, geoscientists can integrate, interrogate, and analyze data to make decisions and reduce inherent business risks in exploration.

An effective location intelligence strategy offers consistent, transparent, and secure support for geoscientists who handle and manage data on regional and global teams. Exploration and mining company leaders must have confidence that their prospective assets are securely maintained at all stages of the mining life cycle, which is enabled by a secure system managed at the corporate level.

Operations

GIS technology can help you locate, track, and manage operations and assets anytime, anywhere. Mine operations encompass many disciplines and workflows that must be coordinated into a single decision-making stream to achieve the highest efficiency and ROI. Companies use GIS in mine operations to capture field data, consume live feeds from mine equipment and sensors, capture and analyze imagery, and monitor mine infrastructure—all integrated in a single near-real-time system.

Effectively planning a blast to fracture rock or other surface material starts with understanding conditions in the field. Using GIS field maps, crews can accurately log and place drill hole locations and collect pre- and post-blast information. Crews can link this data in real time to the mine-planning engineer.

Because the condition of haul roads (roads designed to carry heavy truckloads of materials) can change dynamically in a day, mining crews must have the ability to track road conditions to prevent delays. Through real-time monitoring, GIS can tell operators where to dispatch crews and resources to maintain optimal road conditions, resulting in increased ROI in material movement and reduced equipment repair and downtime.

To obtain an optimal extraction rate, mining staff must monitor stockpiles or storage locations of minerals extracted and processed through a technology called heap leaching. Using the visualization and analysis tools available in a GIS, staff can analyze and track the application material rate and type.

Tailings storage facility (TSF) management is a key aspect of a mine's daily workflow. A TSF consists of at least one dam designed to store uneconomical ore—such as ground-up rock, sand, and silt—and water from the milling process, according to the Mining Association of British Columbia. Milling is a step that involves crushing ore to remove rock and other material before smelting, which combines heat and a chemical reducing agent to produce the metal. Thousands of TSFs hold billions of tons of mining by-products globally. Mine operators are investing in new geospatial and other technologies to manage their tailings more effectively.

Environmental health and safety

Infusing location intelligence into daily operations can help mine operators optimize environmental health and safety (EHS) risk management and overcome logistical challenges. Location intelligence helps mine operators collect and monitor data such as reseeding efforts; bond release schedules; spill prevention, control, and countermeasure (SPCC) tank inspections; and vendor backfilling operations. Once the information is tracked, GIS supports data sharing throughout a mining organization for streamlined communication and record keeping that meet EHS guidelines.

GIS also lets you visualize environmental monitoring data for air quality, groundwater, and more. You can explore historic trends in sampling data to identify potential issues while maintaining a comprehensive and reliable system of record to ensure environmental regulatory compliance.

In mining operations, converting pen-and-paper-based inspection workflows to digital formats saves time and improves accurate reporting. By visualizing and monitoring inspection data, mining staff reduce risk and ensure environmental compliance. Automated report creation, email notifications, and alerts lead to efficient record keeping and quick response times.

From mine closure to reclamation and reseeding, GIS technology, including field mapping and drone support, enables the bond release process, in which government regulators conclude that a mining operation has completed mine reclamation work and so return a required bond.

Logistics

Many mining companies use geospatial technology to meet their logistic needs. Staff use drones to accurately manage inventory volumes. They also identify the best routes for fleets of delivery vehicles to reduce drive and idle times and increase profit. With GIS, managers can remotely monitor real-time operations to identify opportunities for improved efficiency. Promoting logistical efficiency also increases safety and sustainability throughout the global supply chain.

In supporting stockpile and inventory management, GIS can be used to manage materials and resources to keep operations moving forward while calculating stockpile volumes for accurate business planning. Some mining organizations manage complex global and regional supply chains that span many geographies and transportation modes. GIS can help staff map and manage the entire supply chain.

GIS enables market analysis by evaluating business and demographic spatial data to identify expansion opportunities. For example, understanding the market potential of aggregate mining is one key to a successful mining operation. Aggregate mining is the process of recycling material such as crushed rock, sand, and clay for use in asphalt and concrete to create infrastructure such as roads and dams. GIS also helps identify competitor locations and driving times to understand competition and fill gaps in the market.

GIS in action

Next, we'll look at real-life stories that illustrate the successful use of GIS in mineral resource development.

SEEKING THE GOLD STANDARD FOR SUSTAINABLE MINING

Newmont

NOT LONG AGO, NEWMONT, THE WORLD'S LARGEST producer of gold, faced corporate challenges that could affect all aspects of its global operations. Across the business world, companies were increasingly focused on environmental, social, and governance (ESG) issues. Newmont, which initiated a corporate ESG program many years ago, wanted to further distinguish itself as an ESG leader in the natural resource industry.

The company asked these questions:

- How could it more effectively manage and monitor waste products, called tailings, created from the processing of ore?
- How could it meet this challenge without impairing output, damaging the environment, or significantly affecting the bottom line?

"Because effective management of our liabilities is what will differentiate mining companies moving forward, the more we raise our game, the more influence on other producers to do the same," said Mark Casper, senior vice president for resource evaluation and mine planning across Newmont's operations in North America, South America, Australia, and Africa.

New tools for the ancient work of mining

Tailings consist of a slurry of ground rock and chemical effluents that are generated in the process of separating a precious commodity from ore. The slurry is deposited in an outdoor tailings storage facility, or TSF, a containment area like a dam that is often surrounded by earthen embankments and engineered to be structurally and environmentally safe.

Newmont uses GIS technology to locate mineral resources, much as other companies use GIS to scout for energy deposits or plan wind farms. Once a TSF has been constructed, operations managers typically use GIS-based software to monitor on-site sensors and track the facility's performance as part of a broader monitoring program.

The company's commitment to improving the sustainability of tailings management partly acknowledges its significance inside and outside the company. "It touches the communities [where Newmont works]," Casper said. "It touches the environment; it touches the rest of the business. [And] occasionally, we may need to make trade-offs that may be slightly detrimental to costs and performance in the operation to make sure we're doing the right things on an ESG platform."

A few years ago, Newmont needed a global leader to oversee and refine tailings management and further strengthen risk management practices to ensure the safety of Newmont's sites and neighboring communities. The company turned to Kimberly Morrison, a geotechnical engineer and former consultant with significant experience in mining operations. She had the skills and experience to manage complex internal discussions, negotiate corporate networks, and raise expectations without alienating colleagues.

The right person at the right time to elevate tailings management

As Newmont's senior director for global tailings management, Morrison set out to strengthen the methods and standards for dealing with ore-processing waste, improve monitoring using GIS and location intelligence, and communicate the need for those efforts throughout the mining industry.

"We needed somebody that could really [raise] the profile of tailings design and management within the business," Casper said, adding that Morrison was hired to ensure that Newmont attains "the next level" of practice in tailings management.

Morrison's vision was to create a digital twin of each facility—a real-time network of sensors that managers could monitor on GIS maps. The sensors are designed to monitor embankments and dams for changing conditions, track water levels and deformation, and send alerts if readings reach prescribed trigger levels.

To increase Newmont's global visibility in tailings management, Morrison turned to GIS dashboards to track the status of the facilities Newmont oversees internationally. The upgraded system was designed to incorporate drones, some equipped with thermal imaging to detect conditions in earthen embankments that can lead to seepage before it is visible to the human eye.

"My desire is to have visibility to facility performance at my fingertips for any of our worldwide facilities and at any point in time," Morrison said. "We're currently in the process of expanding our use of real-time monitoring of our facilities, and we're working to explore linkage of the drone data with the data from instrumentation and remote sensing sources to support development of the desired four-dimensional view."

The fourth dimension is time. In this case, that includes the past performance of tailings containment areas. Morrison learned the value of using historical data from a GIS specialist more than a decade ago. She saw how data could be transformed and compiled in a map-based platform to provide a complete picture of a site's characteristics. The ability to visualize the data this way supported the federal permitting process to expand a US tailings facility.

Morrison has applied that knowledge to her current work. "We have developed and are in the process of implementing new standards and guidelines for tailings management within Newmont," she said. "We are implementing training programs and workshops to support competency building of our personnel."

In a fast-changing field that demands safety and environmental stewardship, it's important to ensure everyone stays current. "We implemented monthly global team meetings to keep our sites up to speed with new developments, while providing a platform for learning from each other," Morrison said.

She and the team have implemented a web-based app for monthly performance reports of critical control performance at tailings facilities—a report that reaches the highest levels of the organization. "I have seen a paradigm shift within the company with respect to tailings management," she explained.

Morrison, who was named one of 100 Global Inspirational Women in Mining (WIM100) in 2020, has tried to find people outside the company who also are driven to increase the sustainability of tailings management. Casper described her as "passionate about spreading that influence to the broader industry." She is active in industry consortia, including the Society for Mining, Metallurgy & Exploration (SME). There, she is the founding chair of the Tailings and Mine Waste Committee, leading development of a comprehensive tailings management handbook, an external tailings information portal, and webinars and programs. She also serves on the tailings working group of the International Council on Mining and Metals (ICMM) and was an active contributor to ICMM's *Tailings Management: Good Practice Guide.*

Morrison believes the mining industry and the next generation of operational leaders are taking notice.

"With the GISTM [Global Industry Standard on Tailings Management] that spells out certain requirements around governance and roles and responsibilities, the future of tailings management for young professionals trying to figure out what they want to do is very, very bright," she said.

Women in STEM taking on mining's environmental challenges

Briana Gunn, group executive of environmental affairs at Newmont, has been a long-term advocate of Morrison's work. Both work in a science, technology, engineering, and math (STEM) field dealing with vital but challenging environmental issues.

Gunn credits Morrison's success at the company to her firsthand experience at mine sites and her clarity of purpose. Already, Gunn said, she has seen the results of Morrison's dedication. "We were doing tailings management before she got there, but not at as high a level as we're doing now... and not elevating and communicating the important issues throughout the organization."

Along with technical know-how and personal energy, Morrison has another essential element that wins supporters: her integrity.

"It's her ability to look at a situation and say, 'I think that we can do better, and we need to improve our systems,'" Gunn said. "And her willingness and her drive to do that is what I would call her integrity."

That drive, combined with an appreciation for the role of GIS and other technologies, has made an impact at Newmont.

"We needed a real hard look taken at our governance around [tailings]," Gunn said. "It was extremely valuable to have somebody take a look at it and say, 'Let's figure out how we can better integrate risk management into our design, into our governance, into the way we evaluate projects.' And while there's still work to do, she's done a lot in the two years since she started."

A version of this story by Geoff Wade originally appeared in *WhereNext* on April 29, 2021.

ENHANCING OPERATIONS WITH REAL-TIME INFORMATION

PT Freeport Indonesia

THE OWNER AND OPERATOR OF INDONESIA'S LARGEST copper and gold mines, PT Freeport Indonesia (PTFI), is increasing operational efficiency, improving occupational safety, and making more informed decisions with the help of an enterprise-wide spatial data platform. The new data management system uses GIS technology to integrate exploration, operational, geohazard, and environmental data and other relevant information from across the business. The data is then visualized and analyzed on an interactive mapping platform that provides decision-makers with a comprehensive view of their operations.

Satellite image of mining operations overlaid with real-time locations of mining equipment.

Web GIS features real-time data collection, analysis, and reporting capabilities that provide PTFI managers with an accurate, up-to-the-minute understanding of their operations, enabling more precise decision-making.

Having a spatially enabled view of its operations has given PTFI more oversight over its business. Management knows what is happening on the ground in near real time, ensuring effective management of operations and resources.

Consolidating data from all business divisions onto a single platform improves the flow of communication between departments, which also improves collaboration and corporate governance.

The adoption of GIS technology enterprise-wide has helped in these specific areas:

- **Hazard monitoring and ensuring a safe working environment:** The ability to monitor the movement of equipment (trucks, shovels, loaders, and so on) and geotechnical indicators (for example, shifts in the ground, soil settlement, or landslides) simultaneously helps identify and prevent hazards. Real-time reporting enables managers to improve staff safety, avoid accidents, and reduce downtime from incidents. Reducing the number of accidents and near misses creates a more socially responsible operation and allows the company to direct its resources to mineral exploration and production.

- **Corporate governance:** Real-time monitoring of field activities supports management of mining operations. Accurate and authoritative data informs decision-making, allowing effective use of resources, saving time and money.

- **More efficient operations:** Web GIS improves cost accountability and enables more efficient management of mining operations. The system allows monitoring of mining activities related to operations risk management, mine productivity, asset management, and environmental management.
- **Compliance with environmental policy:** Visually representing spatial models and reports allows PTFI to clearly document environmental compliance for government stakeholders and facilitates environmental auditing. Real-time data improves decision-making, enabling potential environmental incidents to be identified and mitigated before they become an issue. This also increases accountability.

PTFI's adoption of Web GIS allows more users to access the system from mobile devices that can input production data on-site in real time, further increasing data accuracy and operational efficiency.

A version of this story by Esri Indonesia titled "Enhance Your Operations with Real-Time Information" originally appeared on esriindonesia.co.id.

A SEA OF OPPORTUNITY IN THE CALIFORNIA DESERT

Esri

CALIFORNIA'S PLAN TO HAVE ALL NEW CARS SOLD IN THE state operate emission free by 2035 could potentially hasten the country's transition to electric cars using lithium-ion batteries. Miners, manufacturers, and logistics companies stand to benefit if the plan increases development of the state's lithium reserves.

If these industry players create a Southern California hub for lithium extraction and battery production—an outcome the state is encouraging—location intelligence will play a large role in making the process efficient and collaborative.

From Imperial Valley to Lithium Valley

Most lithium today comes from Australia, China, and the "lithium triangle" that covers parts of Argentina, Chile, and Bolivia. California's Imperial Valley, a desert region in the southeast corner of the state, could contain even larger reserves.

According to a recent study by SRI International, highlighted in *Bloomberg Businessweek,* the magma-heated brine beneath the Salton Sea could annually yield eight times the amount of lithium produced globally in 2019. With the lithium-ion battery market on track to reach $129 billion by 2027—up from $37 billion in 2019—the Salton Sea's lithium could be in high demand.

Geothermal plants already dot the shore of the inland sea, pumping the brine and converting it into turbine-powering steam. Several companies have expressed an interest in extracting lithium before the brine is sent back underground, according to *Businessweek*.

California officials hope manufacturers will build battery factories—and perhaps even plants that assemble electric cars—creating an industry cluster in a region that officials have rebranded as Lithium Valley.

A unified lithium supply chain

This kind of ecosystem could consolidate an important part of the electric car industry in one area, creating supply chain efficiencies for the companies involved with benefits locally and statewide.

Turning an idea into reality will require cooperation among diverse stakeholders, beginning with energy companies, mining interests, local officials, community leaders, and environmental groups—potentially expanding to include battery and car manufacturers.

In other industries, such cooperation has started with a smart map of an area, stored and accessible with GIS technology. Mining companies routinely use GIS to create detailed maps of their sites for everything from exploration to operations. Integrating satellite imagery, remotely sensed lidar data, and point cloud maps created by drones, companies can visualize a project from many perspectives. The data helps companies identify the location of deposits, strategize how to extract them, organize the daily operations of the site, and track the environmental impact of these activities.

For any cluster that might develop near the Salton Sea, a smart map could be a platform for communications among the various interests, with cloud-based permissions governing which companies and individuals can access which data. That kind of shared ground truth promotes transparency and can drive collaboration within industry coalitions.

Easing the environmental footprint of lithium

One consistent criticism of electric cars as clean-energy solutions is that extracting and processing the key elements in lithium-ion batteries is a carbon-intensive process. Extracting lithium from brine involves large amounts of water. The process requires and produces chemicals that are potentially harmful to air, water, and soil.

Some projects in South America's lithium triangle have received criticism for misuse of local water reserves and for inattention to the environmental cost of lithium mining.

Questions remain about how location intelligence and the use of GIS might help companies locate and build Salton Sea projects that are safe, sustainable, and efficient at an inland sea where migratory birds take refuge and in a region with a tourist industry and growing population. It's possible that GIS technology could monitor the year-long lithium refinement process to identify potential hazards. As lithium production increases, GIS could help lithium-ion battery makers plan factory sites and manage the logistics of distribution.

Growing Lithium Valley

If battery factories materialize, they'll integrate locally produced lithium with other battery components sourced from around the world. Factory owners can use location intelligence to monitor their global supply chain and reduce the risk of disruption, as automaker General Motors (GM) has done for years.

Cobalt, another key component in electric vehicle (EV) batteries, comes from several countries, such as the Democratic Republic of the Congo, that have been accused of exploiting mining labor. To ensure that they do not perpetuate bad labor practices, battery makers could follow the example of food and beverage companies that use GPS, dashboards, and other methods to track the sources of palm oil to control deforestation.

Once lithium in the batteries leaves Imperial Valley, a Salton Sea lithium industry could continue to exert a positive influence. Companies around the world are beginning to see their products as part of a circular economy rather than one that extends in a straight line from the manufacturer to the consumer to the landfill. When Salton Sea lithium leaves the area in battery form to power electric cars around the world, manufacturers could take a location-intelligent approach to recycling. A manufacturer could estimate the life span of any given battery, contact the owner around the time the battery is likely to expire, and offer buybacks. Instead of going to landfills, some components of the battery could be reused and reintegrated into the industrial process, creating a profitable and sustainable circular economy.

Building a lithium industry in the Imperial Valley will take time, determination, and patience. But with a holistic application of location intelligence technology, an outpost of the world's clean-energy future could bloom in California's desert.

A version of this story by Geoff Wade titled "California Eyes Sea of Opportunity for EVs" originally appeared in *WhereNext* on January 19, 2021.

ENSURING THAT MINING OPERATES RESPONSIBLY AND EFFICIENTLY

New Mexico Energy, Minerals, and Natural Resources Department

NEW MEXICO, AT 121,000 SQUARE MILES, IS THE FIFTH-largest state in the United States. Overseeing mining operations throughout such a large area while staying within budget is no small achievement. But the New Mexico Energy, Minerals, and Natural Resources Department's Mining and Minerals Division (MMD) is resourceful and supplements its operating costs by using GIS, acquiring data at no or low cost, and forming geospatial and technical data-sharing partnerships.

MMD ensures that mining operations, from exploration to reclamation, are conducted responsibly. Maps provide a baseline for analyzing activities and disturbances made by mining operations across the state's vast landscape. The agency uses GIS software to process mining operations and exploration permit applications and to report economic impacts.

Two of the four MMD programs, the Coal Mine Reclamation Program and Abandoned Mine Land Program, were created as part of the Surface Mining Control and Reclamation Act (SMCRA) of 1977, which formed partnership arrangements with the US Department of the Interior's Office of Surface Mining Reclamation and Enforcement (OSMRE). Grants provide MMD staff with the funding needed to collect geospatial information. The Coal Mine Reclamation Program developed relationships with mining operators to share maps and geospatial data. Staff from the Mining Act Reclamation Program collect permit locations and track reclamation, using GPS to populate a geospatial database.

To assess mining reclamation operations, the state uses digital elevation models (DEMs) and digital terrain models (DTMs) from various periods, most of which were created by the US Geological Survey. In recent years, MMD has used orthoimagery acquisitions from the US Department of Agriculture's National Agriculture Imagery Program to create newer statewide DEMs and DTMs.

DEMs and DTMs offer geographic representations of what areas were like before mining. The models reveal stock tanks and dams along drainage corridors and identify those that have been breached. The models also depict patterns in drainage basins that indicate sinuosity, which provides a method for analyzing hydrological and topological characteristics. Using those models, MMD asks mine operators to reproduce the degree of sinuosity that's been determined.

GIS helps MMD track mining activity throughout the state and enforce reclamation regulations for surface and abandoned mines. The agency indicates land change, maps mine impacts, provides guidelines for mine reclamation projects, and tracks mandated environmental and cultural assessments before project design.

The New Mexico Environment Department also has a stake in ensuring that mines follow established regulations. GIS supports MMD's collaboration with the state Environment Department in considering permit applications. Maps make it easier for the agencies to review and comment on mine permits and closeout plans as well as ensure that environmental standards are included in each application. The Environment Department also works with MMD to monitor mining reclamation activities.

"GIS helps the Mining and Minerals Division prioritize where it should spend money on surface reclamation projects," said Linda S. DeLay, a GIS professional at MMD.

DeLay used GIS to prioritize cleanup activities. She analyzed the location of legacy uranium mines and ranked their priority for reclamation. The basis for ranking each of these mines was its proximity to streams, agricultural sites, urban areas, and wells. DeLay then used a weighted overlay GIS model to map reclamation priorities. She presented the New Mexico Legacy Uranium Mines map and dashboard at MMD and other agency meetings and at a national conference. The map helped decision-makers decide where to allocate resources.

To monitor coal mine reclamation, MMD applied for a grant from the OSMRE Western Region to acquire WorldView-2 satellite imagery. Using this imagery along with ground surveys of vegetation, the remote sensing analyst created vegetation change detection maps for abandoned coal mines as part of a reclamation project near Vermejo Park Ranch, a private nature preserve and guest ranch in northeastern New Mexico and southern Colorado. The reclamation project is helping specialists evaluate revegetation and wetland mitigation.

Geomorphic reclamation activities have included redistributing and burying the coal waste as well as reforming stream channels to a more natural pattern. The geomorphic work aims to prevent waste from reaching watershed drainages. GPS devices attached to earth-moving equipment helped map the new terrain design. The imagery was so detailed that an analyst could determine the number of pinion and juniper trees in the area.

Analysts also used lidar data. Whereas satellite imagery provides a close-up of surface mines, GIS renderings of lidar data offer a detailed 3D perspective. Sophisticated 2D and 3D maps can reveal the condition of an area before mining operations began.

MMD also used GIS to assess the impact on the vegetation and terrain around the El Segundo coal mining operation and proposed mine in northwest New Mexico. To document baseline landform conditions, the New Mexico Energy, Minerals, and Natural

Resources Department acquired two 25-square-mile areas of lidar data that had been captured before mining operations. Staff used first-return lidar data to render vegetation density images and bare-earth lidar data to model the terrain. The lidar rendering is a blueprint for what the mining company must do to restore the terrain's original contour and reestablish vegetation in its original condition.

Staff found a more affordable way to capture data by using an unmanned aerial vehicle (UAV). A compact camera mounted to a Trimble UX5 fixed-wing UAV takes many overlapping orthophotos. Photogrammetric processing of these images generates a point cloud of x,y,z values rendering a 3D topographic model. MMD used this technique and GIS to create a topographic model of a stream

A camera mounted on a UAV rapidly captures images that go through photogrammetric processing to generate a point cloud for a 3D representation of a stream.

restoration project at a historic coal mining town. By attributing the points with the photo red, green, and blue (RGB) values, staff calculated the heights of the vegetation. To spot-check the accuracy of the remote sensing data, staff went into the field to measure the heights of a sample of vegetation for comparison.

MMD currently uses Microsoft SQL Server, integrated with ArcGIS Server, to manage most of the mine information, geodatabase, and web map applications. The division is transitioning more of its geospatial data to the ArcGIS system. The geodatabase includes data from the state's resource GIS clearinghouse, mine operators, and elsewhere, as well as data generated in-house.

MMD makes mining information available to the public through its website. Users can see the locations and names of mines, which are coded as active mines, inactive mines, and mines where the bond has been released. They can see coal mine permit boundaries, coal districts in New Mexico, and US coalfields characterized by coal type. The map also has a soil type layer and geologic period layers.

A version of this story by Barbara Leigh Shields titled "GIS Ensures That Mining Operates Responsibly and Efficiently" originally appeared on esri.com.

PART 4

TRANSITIONING ENERGY

FORTUNE 500 COMPANY LEADERS ARE COUNTING ON digital transformation for the future success of their businesses. This reliance is widespread in the energy industry, where geospatial technologies—combined with data-driven insights—are transitioning operations, increasing efficiency, and supporting strategic decision-making along the broad value chain. Business leaders also believe that digital transformation can contribute to the collective goal of long-term energy sustainability for the world.

Location data and analytics underpin every element of the energy industry, including finding and developing resources, refining and transporting products, and sales and marketing. Understanding and integrating location intelligence across these enterprise systems propels more insights, drives intelligent operations, and delivers a safer and more sustainable business.

Next, we'll take a detailed look at how businesses use GIS in upstream production and operations; midstream production and transportation; downstream refining and distribution; and health, safety, and environment (HSE) considerations.

Upstream production and operations

GIS supports the planning and design phases of oil and gas businesses in the process of upstream production (the identification, extraction, and production of raw materials). GIS allows for real-time analysis, visualization, and integration of massive volumes of data to perform repeatable workflows and analysis and enable data-driven operations. By bringing together diverse data, including exploration data, GIS also helps you perform quantitative analysis for consistent investment decisions.

GIS can support energy companies at project, regional, and corporate levels. With GIS, web and map services offer many capabilities that allow for direct use and analysis, including the following:

- Manage operational complexities from the office or a mobile device, including wells, pipelines, power and communications, field roads, fences, facilities, people, vehicles, and equipment.
- Bring real-time data feeds together with physical assets to generate a digital twin of production processes to understand operations, model alternatives, and design and implement the most effective options.
- Design and optimize unconventional infrastructure, roads, pads, and well placement.
- Ensure safe and compliant operations with real-time asset tracking and workforce coordination.
- Identify the most effective locations to introduce large-scale carbon capture and sequestration projects.
- Understand carbon dioxide and methane sources, subsurface storage capacity, and available transport infrastructure.
- Evaluate potential technical projects together with economics, regulations, political and social variables, and HSE impacts.

Midstream production and transportation

Because energy companies move large volumes of resources in a dynamic global environment, stakeholders must know what is happening and where in real time. Midstream digitalization combines engineering, real-time business, and location data in a single dashboard, giving leaders a new level of end-to-end operational awareness and analysis. GIS offers an array of capabilities for midstream operations that store, process, and transport products:

- Use maps and apps for reviewing a complete energy network to improve efficiency, prevent disruptions, and gain competitive advantage.
- Fuse location and operational intelligence through user-friendly dashboards and workflows to accelerate inspections, routing, and compliance.
- Ensure a safe and reliable transportation system for products.
- Use smart maps to plan, manage, and maintain fleets, routes, and delivery schedules.
- Use global automatic identification system (AIS) data and big data analytics to reveal trends.
- Optimize shipping through predictive modeling.

Downstream refining and distribution

Maps and spatial analytics can optimize downstream operations that turn materials into final products and sell them to consumers. GIS helps downstream businesses manage refinery operations, improve distribution networks, increase returns across supply and trading, and grow retail networks. Location intelligence leads to smarter decisions and safer and more efficient operations, ultimately improving customer service and increasing the bottom line.

Downstream operators use GIS in these and other ways through the production process:

- Support facility plans based on computer-aided design/building information modeling (CAD/BIM), asset management, emergency response, and HSE reporting.
- Manage work orders, collect field data, and track workers and equipment through mobile mapping apps.
- Create a digital twin of your assets, bringing together operational geography with your local engineering data.
- Explore energy market data to identify opportunities.
- Bring together variables such as weather, news, geopolitical, shipping, pipeline infrastructure, production, and storage capacity to identify trends, understand product movements, and make more informed decisions.
- Support demographic analysis, site selection, environmental monitoring, and fuel delivery optimization.
- Assess market conditions for current and proposed stores.

Health, safety, and environment

The best approach to HSE in the oil and gas industry is prevention. Mapping a path to preparedness with plans for evacuations, containment, and mitigation can save lives and property during an emergency. Maps and spatial analysis reveal vulnerabilities that can be addressed. Knowing where people and assets are in real time helps minimize risk if an emergency occurs.

HSE processes generate copious amounts of data. Data originates from incidents, inspections, noise and air quality monitoring, driving logs, and other sources that are time consuming to collect, prone to error, and costly to manage. GIS facilitates digital workflows to improve collection, reduce errors, and drive new analytics and insights, ultimately leading to safer and more efficient operations. During a response effort, emergency teams must be more informed and flexible at all points. Successfully planning for and responding to growing threats requires agility and effective communication. GIS provides configurable tools for situational awareness, rapid impact analysis, deploying resources, and communicating with the public.

GIS has helped many leading companies realize the strategic, operational, and financial benefits of including biodiversity conservation in their decision-making, policies, and operations. Corporate social responsibility, protecting biodiversity, and net-zero commitments require a proactive approach.

The number of companies publicly committing to reducing carbon emissions has become a barometer for corporate awareness regarding climate change. Business leaders are rewriting strategies around sustainability and employing GIS to understand how their operations contribute to greenhouse gas production and associated carbon emissions. From energy transition and emissions reporting to racial equity and social responsibility, GIS provides data management, analytics, and communications to engage with stakeholders internally and externally through maps, apps, and reports.

GIS in action

Next, we'll look at some real-life stories about how leading organizations use GIS to meet the energy industry challenges of the 21st century.

WITH A COMPREHENSIVE GEODATABASE, OIL PRODUCTION BECOMES MORE EFFICIENT

Kuwait Oil Company

KUWAIT'S OIL ERA BEGAN ON FEBRUARY 22, 1938, WHEN oil was first discovered in Burgan field in the country's southeastern desert. The very first well, named Burgan No. 1, is still producing. Over the last 80 years, scores of other oil fields have been discovered across the country. Today, the Kuwait Oil Company (KOC)—whose primary role is to explore, develop, and produce hydrocarbons within the State of Kuwait—has thousands of wells and more planned to meet future production targets set in its 2040 Strategic Plan.

At KOC, oil production begins at the well and extends through flow lines to preliminary processing facilities known as gathering centers. To get the crude out of the ground and to the 30 gathering facilities throughout Kuwait requires a profusion of infrastructure—cables, pipelines, communication lines, rig roads, fences, and shelters—both above and below ground.

But a few decades ago, the oil field infrastructure in Kuwait became congested out of historical necessity. Iraq's invasion of neighboring Kuwait to the south in 1990 severely damaged the oil fields. As Iraqi forces retreated, they set fire to more than 600 wells and destroyed 10 gathering facilities. After liberation, Kuwait needed to rebuild, and the fastest way to obtain the revenue was to accelerate oil production. Given the scale of the damage, however, and the need to build facilities and ramp up production quickly, this extensive development project ended up introducing inefficiencies into the production process. KOC had to construct new pipelines, flow lines, and power lines around the oil lakes that formed when its facilities were damaged, so new development couldn't follow the existing pipeline corridor. Thus, the amount of usable land on the oil fields was reduced, and subsurface development, including both exploratory drilling and drilling for production, was limited.

To address this congestion, the operations technical services team at KOC formed an Infrastructure Master Plan (IMP) in 2006 that aims to decrease field congestion; keep track of land reservations and resolve conflicts; mitigate land encroachment; and comply with HSE regulations. The team uses GIS to administer all this, specifically relying on the ArcGIS Platform to optimize future reservations for wells, pipeline routes, facilities, and more.

A new way to collect, develop, and implement data

KOC started developing the IMP by building a comprehensive and robust ArcGIS geodatabase that includes detailed geospatial and nonspatial information about all existing and proposed KOC assets. Initially, the IMP team—which consists of GIS specialists, GIS developers, GIS data management engineers, and AutoCAD drafters—designed the geodatabase so it featured a unified data model of all KOC assets. This required employing entity-relationship modeling to ensure that the conceptual representation of the structured data was well designed. After doing a thorough analysis of the model, the IMP team produced a database schema to define and describe the contents of the geodatabase.

Next, the IMP team developed a highly integrated data collection technique for mapping assets that involves doing a topographic field survey of all KOC operational areas. Survey crews are equipped with survey-grade GPS receivers, as well as radio detectors and ground-penetrating radar devices to trace both conductive and nonconductive underground features. Not only does this setup provide a controlled and reliable mapping solution, but it also ensures that the data conforms to international standards of spatial data accuracy.

All known oil features are mapped, including aboveground wells, pipelines, manifolds, and buildings; belowground pipelines and electrical and communication cables; and natural and human-made surfaces, such as bodies of water, oil lakes, and berms. Numerous feature attributes are recorded as well, such as the information on the metal tags attached to each KOC asset; details about the wells, such as when they were built; the types of crude transported by various pipelines; and pipe diameter.

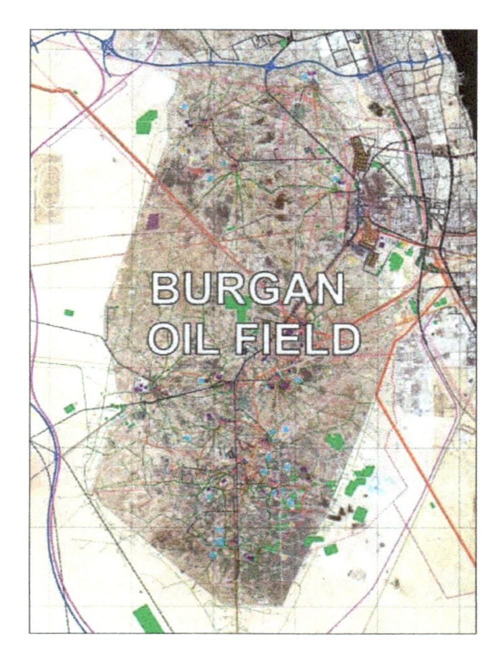

Kuwait's initial oil find was in Burgan field in the country's southeastern desert. Its first oil well, named Burgan No. 1, is still in production.

From there, the surveyors and AutoCAD supervisors on the IMP team process and edit the data in CAD format, and then convert it to a Microsoft SQL Server geodatabase so it can be visualized in ArcGIS Desktop. The data is housed in ArcSDE®, which allows many users to edit the geodatabase synchronously without locking features or duplicating data, thanks to its integral versioning capabilities.

This extensive data collection, development, and implementation effort is giving KOC a large amount of highly accurate and up-to-date geospatial information about Kuwait's oil fields and all the associated infrastructure. Now, this data is serving as the back end for the unique apps that the IMP team's GIS developers are creating to help KOC employees enhance and even expand Kuwait's oil and gas production.

Aligning current needs with future development prospects

Before forming the IMP and adopting GIS, KOC followed an ad hoc site selection process without considering future needs and impacts. But with ArcGIS technology, the company is currently implementing a long-term land management program that will help it secure space for future development, in line with KOC's 2040 Strategic Plan.

Selecting a spot to put a new well is still complicated because of ongoing congestion problems. When a new well is proposed, KOC geologists first determine the desired subsurface location for drilling. When these locations are overlaid on the crowded features, conflicts often exist. The process is further complicated by competing objectives, including costs, practicality, future development plans, operational and maintenance needs, and KOC's rules that set spacing standards between features. To minimize these conflicts, the company has often had to find alternative locations—adjacent to where the oil is instead of right above it, for example—which end up being costlier.

Now, ArcGIS Desktop streamlines the site selection process

by employing geoprocessing tools and frameworks that help KOC secure quality locations for future wells, services, and facilities; optimize locations for better land management; and ensure compliance with environmental health and safety standards. Moreover, the operations technical services team relies on ArcGIS Desktop to evaluate KOC's existing infrastructure layouts and propose cost-effective scenarios for reorganizing assets to ensure that land is used efficiently and economically in the future.

Currently, the map document of KOC's assets is published to ArcGIS Enterprise, where KOC users—ranging from field operators to senior engineers and team leaders—can access and employ it to help ensure the long-term success of land planning.

Continued enhancements with 3D, centralized apps, and real-time feeds

To continue generating cost-effective approaches for land development, the IMP team plans to keep enhancing and centralizing its existing geodatabase and building apps to disseminate this vital information throughout the company. This will facilitate more collaboration and data sharing among teams, which will empower KOC's leadership to make better decisions in a timely manner.

Additionally, KOC uses digital aerial imagery and lidar data from Esri partner The Sanborn Map Company to create high-resolution, natural-color orthophotos and a digital terrain dataset that it can use to make 3D simulations of pipelines; conduct 3D spatial analyses of pipelines, gathering centers, and difficult-to-reach or dangerous areas; perform hydrologic modeling to identify potential as well as unproductive water-prolific areas; and better represent the terrain of oil fields overall. This remote data collection, which can be done quickly and on a more massive scale than survey-based data collection, not only reduces costs but also minimizes risks to people having to map inaccessible, hazardous, and remote areas.

As part of the Infrastructure Master Plan, survey crews collect data on oil production assets and verify features in the field.

KOC is centralizing its app development as well. Instead of making apps one at a time, on an as-needed basis, the IMP team will put them on a web portal where key company personnel can access the latest, most accurate information whenever they need it. This will enable users to find the maps and apps they want, enhance collaboration among teams, and make it easier to share GIS services and geoprocessing tools. The portal will also offer KOC's attribute data via a unified, interactive, map-based interface. Other specialized GIS apps developed by KOC will support the needs of individual teams, including the crisis management group, the environmental health and safety division, and the team that optimizes pipeline routes.

Likewise, the operations technical services team is continuing to integrate existing services that base asset management on real-time information. Having instant access to live data feeds about the weather, for example, gives decision-makers at KOC more context about the environmental issues that continually challenge Kuwait's desert-based oil production processes.

All this development is designed to place the most valuable

geospatial data in the hands of KOC's decision-makers via easy-to-understand maps. But the success of these map apps depends on first having an accurate, comprehensive, and up-to-date geodatabase. And KOC finally has that.

A version of this story by Majeed Al-Muwail, Faisal Al-Bous, Nasir Osman, Fawzi Abdulrahman, Faisal Shah, Ahmed Saad, and Scott Pezanowski titled "With a Comprehensive Geodatabase, Oil Production in Kuwait Gets More Efficient" originally appeared in the Winter 2019 issue of *ArcNews*.

OPTIMIZING FUEL DELIVERIES USING REAL-TIME GIS

Pan American Energy Group

ARGENTINA IS AMONG THE TOP 30 CRUDE OIL PRODUCERS in the world. The country's Pan American Energy (PAE) Group is the leading private-sector energy producer in the southern region of South America. It's an integrated oil company that manages upstream, midstream, and downstream operations, generates electric energy, and is part of the renewable energy sector.

To achieve greater efficiency among its vast operations, leaders at the company decided to optimize PAE's fuel delivery services. These go to retail petrol stations and other customers, such as airports that distribute jet fuel, located throughout the country.

"Argentina is a large country with a varied terrain, so geography plays a big part in efficient fuel delivery," said Gonzalo Fernandez Grossi, who served as senior IT project manager at PAE. "Since improving fuel delivery services is a geospatial problem, we decided to use our ArcGIS expertise to solve it."

A comprehensive fuel distribution platform

PAE has used Esri technology for 15 years. So, for this project, the company relied heavily on ArcGIS Enterprise and ArcGIS Online.

"We have more than 1,000 customers that are serviced by about 280 tanker trucks," said Fernandez Grossi. "The fuel distribution to our customers is an outsourced service."

For invoicing and distribution planning, PAE uses software from SAP and other complementary solutions. However, company leaders were concerned that they couldn't monitor and control fuel distribution as it was happening. In addition, PAE was collecting very little transit information.

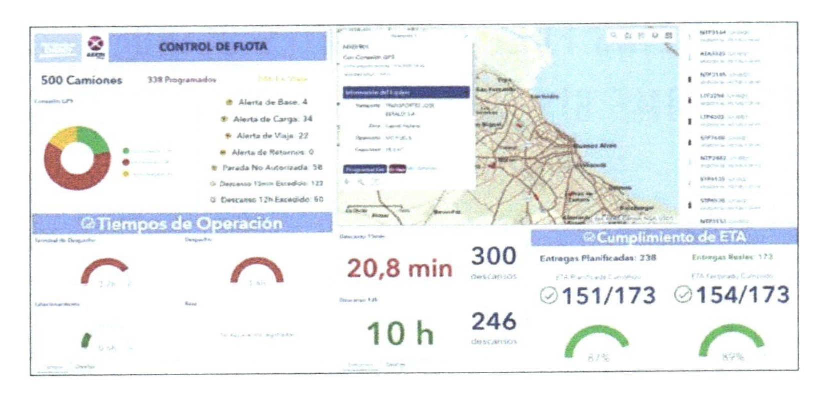

The Pan American Energy Group's new system for optimizing and monitoring fuel distribution benefits staff and customers.

"The theoretical timeline that we originally created provided our customers with a five-hour delivery window, which was very difficult for them from a business perspective," Fernandez Grossi explained. "So, we decided to use Esri software—particularly ArcGIS GeoEvent™ Server—to resolve the lack of complete, accurate, and reliable information during the fuel distribution process. This helped us optimize costs and improve our customer relations and experience."

PAE uses a comprehensive stack of Esri technology in its fuel distribution platform. ArcGIS Enterprise manages connections to PAE's Oracle geodatabase and provides REST endpoints for published data and geoprocesses. The ArcGIS Enterprise portal is the content manager and the system's security management node. This is because it integrates the Lightweight Directory Access Protocol, which is used for directory services authentication. The portal also hosts dashboards and apps that PAE's IT team builds for end users. And ArcGIS Data Store provides an agile and stable repository for storing analysis.

The key to the system is GeoEvent Server, which ingests information—including truck positions, commercial master data, and scheduled deliveries—in real time. It analyzes events, such as fuel drop-offs and driver rest stops, based on business rules, generates alerts, and sends data to ArcGIS Enterprise or Data Store. GeoEvent Server also integrates with email and the WhatsApp messaging service to provide alerts via push notifications.

Another critical component of PAE's new fuel distribution monitoring system is the Origin Destination (OD) Cost Matrix, a ready-to-use service from Esri that can be integrated into apps using ArcGIS REST APIs. The OD Cost Matrix is used to determine the least-cost paths in a transportation network from multiple origins to multiple destinations and to calculate estimated arrival times—in this case, for PAE's fleet of tanker trucks. Additionally, the Node.js web server runs highly complex processes using the OD Cost Matrix to determine the particularities and logic of each trip. Node.js also synchronizes certain data from SAP endpoints (master data and delivery programming) and returns this information to GeoEvent Server for further analysis.

"The system we have developed provides us with complete, accurate, and reliable information during the fuel transportation process and quickly identifies deviations in the transport, allowing us to take corrective actions when necessary," said Fernandez Grossi.

PAE gets more control, and customers get better service

PAE's new fuel distribution monitoring system has proved effective in optimizing its vehicle routing network. It enables staff at the company to oversee the daily distribution of fuels, lubricants, chemicals, and biofuels that are scheduled in SAP. It also allows them to check the availability of hired trucks, monitor how long it takes to load and unload fuel, keep track of travel and driver rest times, and more.

"In the first few months, we determined that more than 50 hours each day were lost due to the micro unavailabilities [small gaps in contractors' work hours] identified. This is the equivalent of losing the use of more than two trucks per day," said Fernandez Grossi. "Now, we obtain better information for the analysis of delays and easily identify optimization opportunities for the trucking fleet."

This includes planning downtime; making improvements to operations at production facilities, service stations, and carrier bases; optimizing loading and unloading processes; and better defining travel times. The system also supports implementing penalties for noncompliance, and it alerts PAE staff when drivers exceed their rest time or make unauthorized stops. In addition, it informs PAE staff when contracted trucks are available to make new deliveries.

One of the best benefits to PAE's customers, according to Fernandez Grossi, is that the system sends them an estimated time of arrival for their deliveries using real-time traffic data alongside the OD Cost Matrix.

"We also send a WhatsApp message to the customer an hour before the arrival time," Fernandez Grossi added.

In the future, PAE plans to extend the solution to its upstream operations and provide greater efficiency and cost optimization for work order programming and monitoring.

"We would also like to promote 'where' as the foundational element in our business," said Fernandez Grossi. "This will include integrating location in our other projects to develop synergy among them. We think that the *where* is a key factor in the implementation of an effective digital transformation for the Pan American Energy Group."

A version of this story titled "Energy Company Optimizes Fuel Deliveries with Real-Time GIS" originally appeared in the Winter 2022 issue of *ArcNews*.

CREATING A COMMON OPERATING PICTURE

Oil Spill Response Limited

OIL SPILLS CAN QUICKLY BECOME ECOLOGICAL AND socioeconomic disasters if not rapidly addressed. Oil Spill Response Limited (OSRL), the largest international industry-funded cooperative, knows all too well the challenges oil spills present. Operating 24/7, 365 days a year, OSRL works tirelessly to provide preparedness, response, and intervention services for potential oil spills across the globe.

As part of an ongoing goal to create a globally efficient response system, OSRL turned internally to find areas for improvement within people, processes, workflows, data systems, and best practices.

The challenge

Although rare, large oil spills can be catastrophic, so OSRL must be always prepared to respond quickly and effectively. If a spill occurs, OSRL decision-makers need a comprehensive picture to ensure deployment of personnel and equipment from any of its 12 global locations to the spill site and response coordination hub within a matter of hours. Many factors must be considered and decisions made quickly regarding environmental, wildlife, and economic impact, protection of ecological resources, inland response, and potential shoreline cleanup. All these issues are factored through OSRL's Emergency Operations Centre (EOC), where decision-makers can gain critical situational awareness and prepare the necessary response measures.

Paper-based fieldwork forms and surveys had resulted in workflow delays transferring the information to the EOC. The old systems also restricted responders' ability to see the full picture from the field. Intending to create both streamlined field processes and better

OSRL staff using paper forms.

situational awareness, OSRL sought to digitize its various field data collection channels.

The solution

Leading the transition from paper forms to digital apps, Liam Harrington-Missin, technology innovation lead at OSRL, sought to ensure that the new system was accessible to everyone and didn't create any additional paperwork in the process.

"What we were effectively doing was removing ourselves as much as possible from pen and paper and moving to a 'digital by default' mentality using tablets, mobile apps, and digital screens," said Harrington-Missin.

Harrington-Missin also wanted to integrate already successful best practices into the new strategy. He started with the widely adopted "toolbox talk"—a previously paper-based process for any task that includes a team briefing to discuss potential risks and preventive measures and identify task-driven goals. Harrington-Missin adapted this toolbox talk paper form into an app-based mobile survey using Survey123.

The digital version provides universal access and feeds data into OSRL's Microsoft Office 365 and ArcGIS Online databases, where it can be transformed into digital dashboards and reports. Survey123 also provides field capabilities to streamline data entry, including audio recording tools, the ability to pinpoint locations on web maps, and hiding irrelevant questions.

"These changes made information faster to file, easier to search, and ultimately [made it] easier to spot trends," noted Harrington-Missin.

Next, Harrington-Missin turned to the issue of aerial surveillance, aware of its impact in the first few days of an oil spill when, typically, little ground reconnaissance information is available.

"Quick aerial surveillance is a high-pressured service that has to work. If it doesn't, not only does it cause responders a great amount of stress to fix it, but it delays critical information needed all the way to the top levels of decision-making," said Harrington-Missin. "So, we are experimenting with ArcGIS QuickCapture and Survey123 as a starting point. Then we bring on ArcGIS and various Esri tools to use them side by side to enable flexible and chained field workflows."

For Harrington-Missin, a factor in successfully updating aerial surveillance information is ensuring the app is easy to use and doesn't distract the observer from the task at hand.

"Our observer needs to be looking out the window more than at the app," he said. "That means the app needs to be incredibly streamlined and easy to use, with big buttons, et cetera."

The results

Harrington-Missin and the team at OSRL are on course to achieve their immediate digitalization goals. Additionally, OSRL's digital changes helped streamline field processes and created a more comprehensive situational awareness. But for Harrington-Missin, this is just one step in a continued global strategy going forward.

"The goal here isn't just to make OSRL's response efforts more efficient—I want to help ensure that the oil response practice as a whole is more efficient," said Harrington-Missin. "This calls for a different strategy requiring greater collaboration across the industry and a movement of GIS experts supporting the oil and gas sector."

OSRL represents around two-thirds of the offshore energy operators in the world, equating to about 170 organizations varying in size, with roughly 80 percent already using Esri technology. Looking toward the future, Harrington-Missin hopes to rekindle conversations around a common operating picture and platform for oil spill response across the industry. The goal moving forward is to work with members of OSRL to share ready-made oil spill tools, provide access to both OSRL and Esri response teams, and create an open and collaborative global common operating picture and platform.

A version of this story titled "Creating a Common Operating Picture with the Help of ArcGIS Survey123" originally appeared on esri.com.

LOCATION INSIGHTS POWER THE SOLAR AND WIND ENERGY INDUSTRY

Aegean Energy Group

WIND AND SOLAR ENERGY DEVELOPERS LOOK TO AEGEAN Energy Group to improve alternative energy construction and operations. The company's Maps to Megawatts solution supports development, on-site analysis, site control, and reporting. Data collection in the field is essential for the company to accurately deliver intelligence throughout the life cycle of its customers' energy projects. Accordingly, Aegean Energy has extended the capability of Maps to Megawatts by enhancing it with GIS tools that streamline field data collection and deepen location analytics.

The challenge

Maps to Megawatts helps clients with planning, strategy, and development. Aegean Energy offers suitability analysis for developers looking for wind and solar energy opportunities and identifies any costly or challenging obstacles to development, such as safety, civil, or environmental issues. Aegean Energy needed a more efficient way to gather data for analysis and make results available promptly to stakeholders.

"Field information is important to us internally," said Woody Duncan, senior vice president of Aegean Energy Group. "It also helps clients be strategic and methodical in project development, which is a time-saver and critical in this day and age. We have to have greater insight as we look for energy opportunities."

Aegean's process for receiving information from the field for analysis needed retooling. The company gathers large amounts of data about a project during construction, but processing that data

The aerial surveillance crew requires an easy-to-use survey app to capture data seamlessly.

was slow. From project photos that required cataloging and filing to paper forms that staff manually entered in an Excel database, the workflow caused a significant lag in distributing information from the field. The company estimated that these slow and burdensome processes took around three months to complete.

Aegean Energy also sought a more efficient way to conduct analysis and quality assurance/quality control (QA/QC) when analyzing data.

The solution

The company implemented ArcGIS Insights℠ to streamline data input, analysis, and reporting. This analytic software allows users to perform data analysis, document their workflow, and share analysis results with others. Aegean Energy uses Insights throughout different stages of the assessment process, including quality control, monitoring ongoing problems in the field, and data reporting.

"We use Insights for quality control, reporting, tracking schedules, and budgets," said Clint Cook, senior vice president of engineering and construction for Aegean Energy Group. "It's exciting because we can now do so much more with data."

Maps to Megawatts functionality includes construction monitoring that tracks inspection status, which is based on data on forms completed by field operations teams. These forms help Aegean Energy identify potential development issues. Once an operator submits a form, it is processed using GIS. Insights flags tasks that are open and ready for inspection. Inspectors have an easy-to-see visual representation that shows where work needs to be done. Using an app to update the task status, inspectors can instantly close the loop, and Insights tracks the change. This simple workflow has improved inspection efficiency.

Aegean Energy also uses the Insights Link Analysis tool to trace form entry errors, another aspect of QA/QC. Using a network of interconnected links, the tool identifies and analyzes relationships that are not easily seen in raw data. Suppose that an on-site worker inaccurately enters two locations on an inspection form. Since each inspection status task links to a specific solar pad number or a wind turbine location, only one form can be entered for each asset. A link chart, with a clear graphic representation, easily shows a pad number with two different statuses, which indicates data entry error.

Duncan noted, "Using this tool saves hours of work. Previously, somebody would have to spend many hours trying to dig through the reports and figure out if information had been entered twice or if there was another issue. But with Insights, we can immediately find the problem. It's like pulling a needle out of a haystack in an instant rather than months later."

Aegean Energy also uses other ArcGIS tools to publish analysis results and expedite reporting. After staff conduct a suitability analysis, they make the results available to web clients or accessible via ArcGIS Online web services. Companies use the results for construction assessments in the field.

Aegean Energy has further increased data capture efficiency using Esri apps—Collector and Survey123. Collector is designed for accurate data collection, whereas Survey123 facilitates digital survey distribution for simplified data collection. Staff use these tools to collect and organize data, quickly connect it to the office, and share it with stakeholders.

The result

"We're using Collector and Survey123, which allows our field folks to fill out all the information in the field and simply press a button to report results," said Duncan. "[To view] that linkage during construction and across multiple disciplines is tremendous. To have the ability to go through an enormous amount of information and simplify it into what we need to see is terrific."

Insights plays a valuable role in ensuring the accuracy and availability of project-related documents:

- On-site workers and administrative staff can sign in to the application and easily view completed forms to ensure correctness.
- Stakeholders can view data and project-related developments, allowing stakeholders—no matter their location—to view data and project-related developments.
- Project managers, senior executives, site administrators, and other stakeholders can see any development issues and whether a project is on time and on budget, accessing web-based data and results available in one location.
- Staff can more easily transmit field data with increased efficiency through improved data collection, eliminating the need to spend hours at the end of the day sorting and sending forms and photos.

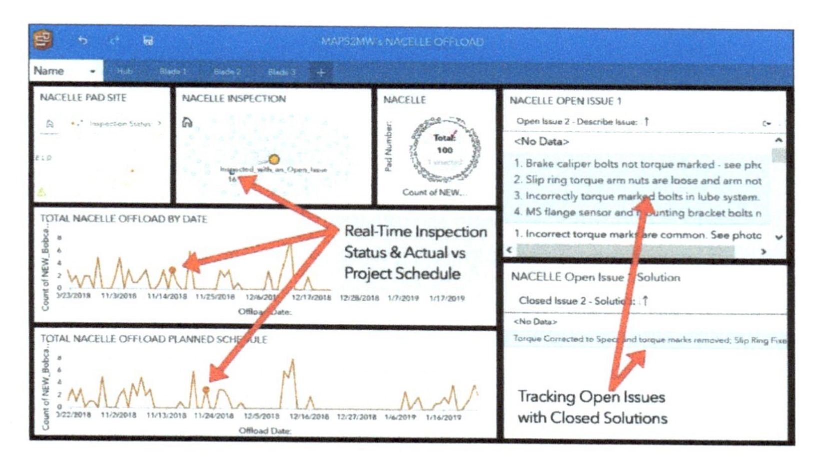

Aegean Energy uses Insights to compare real-time asset inspections with its projected schedule. Additionally, open issues are returned when the user clicks on an asset. This visual analysis saves Aegean Energy hours of work.

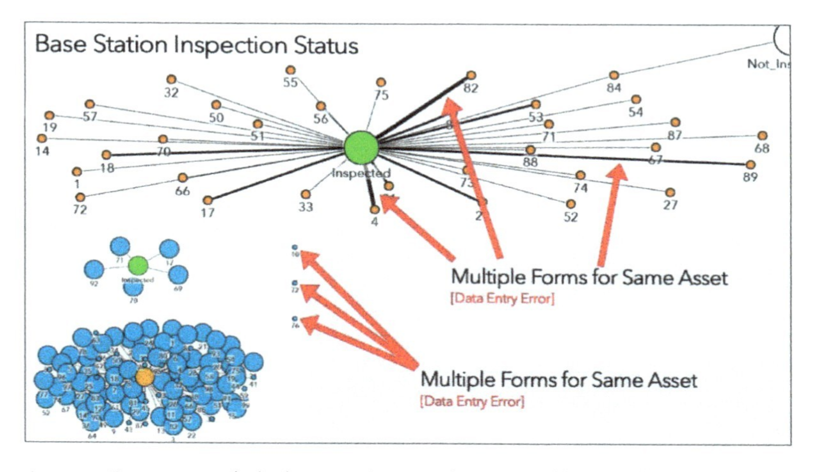

Aegean Energy uses link charts within Insights to quickly identify data entry errors. If an asset has a thick line connected, or an inspection status is not reported, it signifies a data error.

- On-site workers now use tablets and mobile devices to complete forms and quickly share information and results on the web.

The newly streamlined workflows and location-intelligence tools benefit staff in and outside the office. Viewing project issues in Survey123 and tracking task status and resolution have brought great value to customers and the company.

By enhancing the Maps to Megawatts solution with tools from Esri, Aegean Energy Group offers energy companies greater efficiency and understanding for conducting operations and developing construction for wind and solar power.

"For us, we live in maps," said Duncan. "Everything we do is tied to where, so the spatial data we gain from Insights is incredibly valuable; otherwise, it's just data."

A version of this story originally appeared on esri.com in 2019.

PART 5

IMPROVING DELIVERY

THE COST-EFFICIENT MOVEMENT OF NATURAL GAS AND liquid energy resources on land requires a complex network of transmission pipelines. These networks have grown in importance as the world transitions to more sustainable forms of energy. At the same time, the job of pipeline professionals has become increasingly difficult as they strive to meet regulatory and business demands. To address these challenges, pipeline operators use GIS to perform essential tasks and workflows. GIS technology ensures they have the spatial capabilities to meet their current and future business objectives.

Today, pipeline operators use GIS in these and other areas of their work:

- **Gathering pipelines:** Crude oil, or natural gas, is brought from wells to a processing plant, refinery, or transmission pipeline. GIS helps field operators standardize and improve daily processes, such as asset inspections, operational surveillance, and surveys. It also allows them to share data readily.

- **Crude and refined products transmission:** Crude oil is transported to refining centers and distribution terminals, and then the refined product is sold. With location-based information, pipeline operators gain new insights that improve analysis and decision-making throughout the transportation cycle of hazardous liquids.
- **Natural gas transmission:** Natural gas is transported from processing to distribution. Pipeline operators use maps and analytics to access accurate, consistent, and secure pipeline data, using ArcGIS to find, share, and analyze information.
- **Vertically integrated gas utilities:** Gas utilities operating an integrated pipe network own many levels of the supply chain, including transmission and distribution subnetworks. Integrating location intelligence with GIS—using spatial information to support decision-making, prediction, and new insights—drives operations and delivers a safer and more sustainable business.

Within each of these areas, pipeline operators have developed sophisticated workflow-based capabilities to support their common operations. Next, we'll take a more detailed look at how GIS is used in planning and design, asset and integrity management, operations, and health, safety, and environment, or HSE.

Planning design and construction

Pipelines grow through new construction and acquisitions. Pipeline engineers need spatial analysis tools to support planning, design, and construction to integrate different layers of authoritative data into a holistic process. Operators use GIS to select the best path from the origin to the destination—considering the nature of the pipe, its surroundings, environment, and cost of construction. Operators can use digital as-builts (digitized records of completed construction projects) on mobile devices to keep everyone up to date with progress on tasks such as clearing, grading, and right-of-way restoration. Digital as-builts also help operators manage permit applications for their rights-of-way to ensure they are free from vegetation, structures, and other encroachments.

Enterprise asset management

ArcGIS enables improved decision-making and management of pipeline assets through their life cycles. The ability to create and maintain a digital twin of the in-ground pipe network and its surroundings allows users to work with that digital twin as they would with a real network. Once pipes and related components in the trench are covered, a digital twin becomes a pipeline's primary authoritative record of assets in the ground. Most pipeline operators now store their pipe network and GIS data in ArcGIS in Esri's Utility and Pipeline Data Model (UPDM) or Pipeline Open Data Standard (PODS), most recently using PODS 7 or higher, where it can be linked with other enterprise systems.

Integrity management

Pipeline operators must stay vigilant to assess potential actions that could cause their pipe network to lose containment. To prevent such an eventuality, they use sophisticated software to identify, prioritize, assess, repair, and validate the integrity of their pipelines. The relevant workflows cover operating pressures; class locations; high consequence areas (HCAs); and the use of devices called pigs to inspect, maintain, clear, and perform other tasks on pipelines during a process known as pipeline pigging.

Operations

Operating pipelines safely and reliably requires a large workforce collaboratively performing many tasks. Efficient routing, logistics, and real-time dashboards optimize operational performance. Pipeline operators continually patrol and assess the condition of the pipelines from the air and ground and through real-time sensor feeds and imagery. On rare occasions, an incident may occur that demands immediate response, in which case GIS supports risk mitigation, readiness, response, and recovery. Integrated maps in dashboards provide a comprehensive and engaging view of the data needed to make decisions quickly and plan the next steps.

Health, safety, and environment

HSE is of utmost importance to pipeline operators. GIS helps reveal and address conditions that threaten a pipeline's workforce, the community it serves, and the environment in which it operates. Sensors integrated into vehicles and devices used by mobile crews as they perform inspections and maintenance tasks identify and display the location of employees on dashboards in real time. If an employee is in a hazardous area, an alert can be triggered to the employee's location-aware device. Operators rely on GIS to assess the risk of pipeline failure using spill and plume modeling and provide mechanisms to alert the public of unfolding dangers.

GIS in action

Next, we'll explore how pipeline organizations apply GIS for better decision-making and activity management.

IMPROVING EFFICIENCIES WITH LOCATION INTELLIGENCE

Crestwood Equity Partners LP

OIL AND GAS INDUSTRY OPERATIONS ARE DIVIDED INTO three stages: upstream (exploration and production), midstream (processing, storing, transporting, and marketing), and downstream (refining and distribution of by-products). As a midstream company, pipeline operator Crestwood Equity Partners LP transports unprocessed petroleum from upstream oil and gas exploration and production facilities to downstream refineries, where it is processed into various products for distribution and commercial sale.

Crestwood uses ArcGIS to manage and maintain the integrity of its extensive assets as it gathers, stores, and transports crude oil, natural gas, and natural gas liquids. These assets include physical infrastructure, such as compressor and pump stations, pipelines, valves, and storage facilities, as well as real estate rights, including rights-of-way, easements, and surface leases.

In the United States, more than 70 percent of refined and unrefined petroleum products are shipped through 2.4 million miles of pipeline that crisscross the country. These products include crude oil; fuel, such as gasoline, diesel, and jet fuel; and natural gas, both liquefied and vaporized.

The midstream industry has implemented data models to store and analyze the information needed to monitor and maintain petroleum pipeline networks in the United States. Data models maintain construction details, inspection results, integrity management, regulatory compliance, risk analysis, and operational data.

According to Craig Hawkins, former manager of asset information and GIS at Crestwood, midstream companies generally use

Crestwood has implemented data models to store and analyze the information needed to monitor and maintain the millions of miles of petroleum pipeline networks throughout the United States, including this pipeline under construction.

similar methods to store data. Many data models used by midstream companies are cooperatively developed by the pipeline industry with the support and direction of organizations such as the Pipeline Open Data Standard (PODS) Association, which has been developing these models for more than 20 years. Members of the association contribute to the development of the data models and use the latest model available.

The models use a linear referencing system (LRS), which stores the geographic location of a pipeline asset by its relative position along a measured linear distance, rather than determining the coordinates of each asset as other models do. LRS specifies features (or assets) along linear route sections between defined start points and endpoints. It is commonly used by engineers because of its precision. To perform its linear measurements, LRS references the centerline

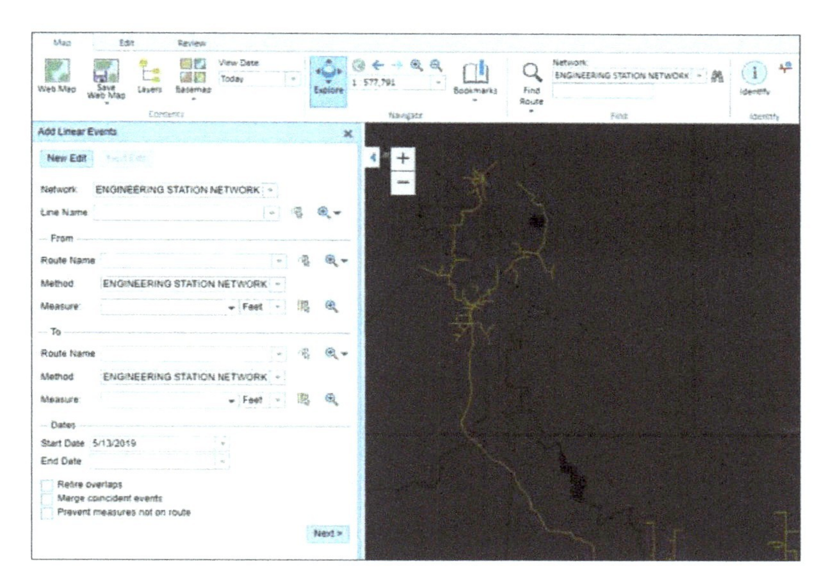

ArcGIS Pipeline Referencing tools provide useful GIS capabilities to perform linear referencing data management.

axis of a pipe, which is a calculated line that runs longitudinally through the midpoint of the pipe's diameter.

In addition to being more precise, LRS is also more cost effective for data maintenance procedures. When a change is made to a section of the pipeline, the affected section is recalculated by the software and stored in its related data model, maintaining the referential integrity of the pipeline. Many pipeline companies use ArcGIS in their operations. Models developed by PODS have provided links to the ArcGIS geodatabase for more than a decade.

After Hawkins joined Crestwood in 2016, he consolidated the company's different data models into a single system. He knew that Crestwood could migrate disparate data sources into a single database so the company could manage its data and perform analyses more efficiently.

Crestwood implemented the extended version of PODS Next Gen, which is supported by the ArcGIS Pipeline Referencing location model and provides GIS capabilities that are useful in performing linear referencing data management. Pipeline Referencing allows the visualization and editing of linear-referenced data in both 2D and 3D.

"Migrating multiple databases at the same time can be complex, but ultimately, we were able to generate one LRS network to get a single, comprehensive view of our pipeline network," Hawkins said. The Pipeline Referencing tools allowed Crestwood to manage centerline geometry and data maintenance daily, he explained. "If a pipeline is rerouted, the centerline and any associated data can be edited," he said.

In addition to streamlining data management best practices, the company is looking at better ways to use geospatial technology, reporting, systems integration, and big data infrastructure.

"I think that geospatial technology holds tremendous potential for our industry because it can help us improve our efficiencies through the use of location-based intelligence, which is already a strategic part of our information infrastructure," Hawkins said.

A version of this story by Jim Baumann titled "Improving Efficiencies in the Pipeline Industry with Location Intelligence" originally appeared in the Summer 2019 issue of *ArcUser*.

MANAGING HYDROCARBON TRANSMISSION PIPELINES WITH GIS DASHBOARDS

Saudi Arabian Oil Company

THE SAUDI ARABIAN OIL COMPANY (SAUDI ARAMCO) operates and maintains an extensive network of hydrocarbon transmission pipelines that carry oil, natural gas, natural gas liquids (NGL), and even products such as diesel. The total length of these pipelines extends almost 15,000 miles.

This network of buried pipelines connects refining facilities, domestic customers, and export terminals. Pipelines cross the entire territory of Saudi Arabia, encompassing an area of diverse terrain and harsh environmental conditions.

Operating the network and its related assets poses many challenges, which often require implementing normal and emergency modifications to maintain the safety, reliability, and integrity of the transmission pipeline network.

The status of changes to the network and its assets must be tracked to ensure they comply with various requirements. The transmission pipeline network, its related assets, and all modifications to the network and assets are maintained as an as-built pipeline network layout monitored by a GIS application. The data for that GIS application is managed by a dedicated unit within Saudi Aramco called the Pipelines Data Management Unit (PDMU).

Because the pipelines and assets are distributed over a wide area, geographic zones or areas of responsibility have been created for the three main subdepartments within Saudi Aramco's Pipelines Department and for their reporting operating units. This structure ensures project completion and compliance for normal and emergency changes.

Project staff come from different disciplines, such as operations, maintenance, and inspection. They submit change requests, update progress data, and confirm the completion and implementation status of changes in the corporate enterprise resource planning (ERP) system.

PDMU staff are not a permanent part of this process. To obtain information and data related to any changes implemented on the transmission pipeline network, PDMU and Pipelines Department staff require intensive email communication. These messages provide Pipelines Department management with status updates.

Developing the Pipelines Management of Change application

To address the challenges of monitoring the pipeline network, a GIS dashboard solution was developed. This solution, the Pipelines Management of Change (MOC), enhances the tracking and monitoring of changes to the network and establishes a complete data update workflow for capturing and maintaining current pipeline asset data.

The MOC application displays information about the completion and compliance of changes to the pipeline network.

MOC is a GIS dashboard with two main components. The first component displays tabular data and change locations on a map. Selectable analysis charts and tables allow managers and stakeholders to browse various types of information about normal and emergency changes. Managers can track the progress of changes and their status.

The dashboard displays status charts or tables on the map georeferenced to its area of responsibility. By browsing charts and tables, you can identify changes that have passed the scheduled implementation date. The distribution of normal and emergency changes is evident because their locations are georeferenced on the map. This visualization allows staff members to arrive at conclusions more quickly and decide on next steps.

The other main component of the GIS dashboard is a data update workflow. Because the ERP system is used by different Pipelines Department staff members for submitting MOC change requests based on geographic areas, this component initially registers any new normal or emergency changes that have been entered in the ERP system each day.

PDMU staff, using the analyses implemented in MOC, can focus on submitted requests for changes—classified as either emergency or normal—that will modify transmission pipelines and related asset data. A location tag on the map is created to reference the MOC request in case it can't be referenced using an existing GIS asset. This step is important for locating new assets that have been added or existing assets that have been relocated.

Based on the status of changes or their completion, a data update workflow is triggered. This workflow is monitored by PDMU, which ensures that the data is updated quickly and passes QA/QC processes. The GIS technology used in this workflow enables users to zoom to asset locations affected by the MOC request and perform

QA/QC on updated data by either approving it or rejecting it and redoing the data update action.

With this solution, staff can identify assets that have been removed or reallocated or newly added assets. Innovative GIS editing tools, email notifications, and QA/QC processes with appropriate levels of review and approval maintain up-to-date pipeline GIS data.

Through a dashboard, MOC employs GIS technology to enhance visualization and track the location of normal and emergency changes submitted by staff members who have different areas of responsibility. The app is integrated with the corporate ERP change management system and email gateway. The solution consists of a data synchronization agent that periodically updates the change request data in ArcGIS Enterprise by importing data from the corporate ERP system through a representational state transfer (REST) endpoint.

MOC was developed with flexibility in mind using ArcGIS API for JavaScript™ and other enterprise technologies. Different user roles have access to specific functionalities such as data updates.

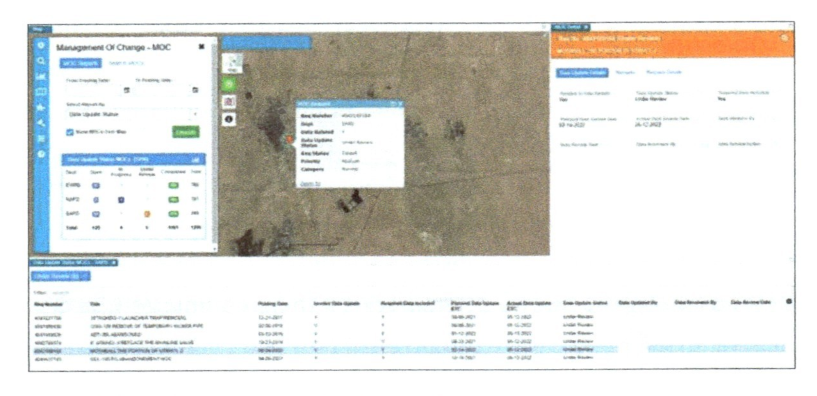

The MOC application shows the data update status of a normal change request.

A designated user role grants data access to PDMU staff. The supervisor role determines who can approve or reject data updates.

Benefits realized

The MOC application provides requested functionality and addresses concerns raised by management and technical staff. The GIS dashboard provides management with selectable charts and tables so the progress of normal and emergency changes can be monitored immediately and compliance can be ensured. These changes are extracted from the ERP system so that MOC acts as a gateway for browsing information without logging in to the system.

MOC enables the classification of changes that involve data updates to the pipeline as-built network and tracks the completion of changes, ensuring data is updated accordingly. PDMU staff can monitor and track all changes implemented on the pipeline. The custom GIS-based workflow helps keep GIS data current. The workflow also eliminates work otherwise needed to customize the corporate ERP system, which directs all pipeline-related changes to PDMU for review. ERP system customization would have delayed the project considerably and imposed major delays in executing changes. MOC solves this problem.

A version of this story by Thamer Tarabzouni, Nawaf Bakr, Iftikhar Ahmed, and Kamran Hussain titled "Hydrocarbon Transmission Pipelines Managed with GIS Dashboard" originally appeared in the Summer 2022 issue of *ArcUser*.

PART 6

GROWING RENEWABLES

GIS SUPPORTS THE DEVELOPMENT OF RENEWABLE ENERGY sources such as wind, solar, hydrogen, and geothermal—and the infrastructure that supports them. Geospatial analysis can optimize energy sourcing and transmission systems and change the way companies use renewable resources.

GIS has an array of uses in renewable energy:

- Use maps, imagery, and remote sensing data to understand energy potential, drive site selection, and improve operational performance.
- Integrate field, machine, and real-time data into operational dashboards that help improve workflow efficiency.
- Maintain clean and smart energy projects and help ensure a low-carbon, sustainable future.
- Deploy systems on-site and in the cloud with modern architectures and software as a service (SaaS), data as a service (DaaS), and infrastructure as a service (IaaS) to improve energy production, transmission, and delivery for greater economic and social resilience.

Renewable energy sources

GIS also enables new energy production by identifying sites of maximum energy potential and optimized economic development while minimizing environmental impact in several ways:

- **Wind energy:** Unite all the information from design to build, and operate wind energy facilities safely and efficiently.
- **Solar energy:** GIS applies across the solar energy business, from mapping energy potential and commercial analytics to engaging with stakeholders.
- **Geothermal energy:** Important GIS-supported workflows include determining potential markets and prime locations to implement geothermal technologies and the required infrastructures.
- **Hydrogen:** GIS modeling is used heavily in designing hydrogen infrastructures to meet demand, to assess market potential, and for resource analysis. GIS allows consideration of several multivariate scenarios to produce optimal business strategies.

GIS can facilitate the collection and management of project data and the analysis of infrastructure placement and restrictions while also reducing business risk in these and other ways:

- **Planning and assessment:** Take advantage of online resource assessment data for project planning, land management and permitting, interactive modeling analysis, and visualization capabilities to build site-specific project plans that will benefit from mobile apps and real-time decision support dashboards.

- **Construction and development:** Manage the complexities of multiple active project developments through powerful 3D design visualizations, and then remotely monitor and share the as-built construction phase with key stakeholders.
- **Operational management:** Effectively manage HSE, profitability and performance, and stakeholder interests through mobile apps, imagery, and IoT sensors for situational awareness, using interactive dashboards. Dashboards bring business intelligence workflows to life and build positive relationships through intuitive communication tools.
- **Policy compliance:** Discover how new policies and regulations affect a company's strategic planning initiatives. Planning new projects and operating them effectively in a rapidly evolving legislative framework requires a clear geographic approach to regulatory compliance, corporate integrity, and transparency for stakeholders.
- **Technology and innovation:** Renewable energy leaders are deploying new and sustainable technologies that drive cost efficiencies, enable safer operations, and provide greater revenue opportunities. Unencumbered by legacy practices, leaders can create a more efficient, sustainable business and future.

GIS in action

Next, we'll explore how organizations use GIS to improve clean energy production, transmission, and delivery; identify renewable energy trends; and optimize analytics.

MAPPING PRIME RENEWABLE ENERGY SITES

Kentucky Energy and Environment Cabinet

THE STATE OF KENTUCKY IS TRANSITIONING FROM BEING A coal powerhouse to becoming a compelling locale for renewable energy generation. Once the leading producer of coal in the United States and still one of the top three coal-producing states, Kentucky envisions a future powered by alternative energy sources—hydropower, biomass, and solar.

Developers are looking for solar power locations as Kentucky faces declining coal use and production along with increasing interest from corporate buyers for renewable resources. With its robust infrastructure and available space, including previously used mine lands, Kentucky bills itself as an ideal choice for solar production sites. But where do they put them?

"For several years, as solar has come down in cost, it's become more of an option here in Kentucky," said Kenya Stump, executive director of the Kentucky Office of Energy Policy. "We had a lot of questions from people: 'It seems like our mine lands would be great for solar,' or 'I don't know why we don't put solar on our mine lands.' And to me, that was always a geospatial question: Where should new sites go?"

Until a few years ago, the Office of Energy Policy had no mechanism to receive and respond to site inquiries, so developers often selected sites with little to no input from the state. "Solar is so new in Kentucky that we had solar developers making siting decisions, but we didn't understand why they were choosing their locations," Stump said. "We would just get notified that a new solar project was going here or going there."

Staff from the Kentucky Cabinet for Economic Development would pass along inquiries to colleagues at the Office of Energy Policy, who, initially, had no way to gauge suitability. Traditional industrial development sites weren't meant for solar, and siting characteristics such as topography, slope, or the presence of threatened species had to be considered. Teams at the Office of Energy Policy used their GIS technology to create the Solar Siting Potential in Kentucky platform, guiding solar developers to prime locations.

A smarter solution for site selection

With input from the state's GIS experts and Esri, Stump and others from the Kentucky Energy and Environment Cabinet applied GIS to conduct site suitability analysis on land parcels available for development. The analysis evaluates sites based on criteria impacting the technical feasibility of construction, and then gives each parcel a score.

Stump's team collaborated with mining consultants and solar developers to collect relevant data layers and determine the criteria needed to compare sites. Support provided by the Kentucky Geography Network (KyGeoNet), the geospatial data clearinghouse for the state, facilitated this collaboration. Melissa Miracle, a former IT consultant who was deeply involved in KyGeoNet, worked closely with the state's nature preserves to include data on local endangered species.

"That was a tough one because we didn't have direct access to the raw data," said Miracle, who now works for Esri. "So we worked with the folks within the agency who handle the nature preserves to create the layers that we needed. It was important for them to be involved so the developers would know where they can and cannot build." Miracle also worked with the Natural Resources division to create a better understanding of the state's mines.

"That was a whole learning curve," Stump said. "The attributes, how they code things, understanding mining reclamation terminology—all that was huge."

Through this collaboration, the team addressed the multifaceted concerns of solar plant providers—including favorable slope, land classification (barren land, mixed forest, and cultivated crops), access to electric transmission lines, population density, proximity to the habitats of threatened species, and status as federal or protected lands.

Communicating with communities

Another benefit of Solar Siting Potential is its transparency to communities that may be affected by possible development. "Our land-use planning is done at the local level. So this tool also helps our local communities understand the reason why developers are looking at their lands—whether it's because they have the right slope or they have access to transmission or other characteristics," Stump said. "It's still up to that community to decide what they want to do with the land."

The platform also helps community stakeholders determine which local land is being considered for solar projects. This information influences personal and local planning decisions, such as safeguarding land for specific uses or protecting views.

"This tool couldn't have come at a better time because it shows stakeholders that we are actively trying to find places that were not prime farmland," Stump said. "We have a layer in the tool that you can turn on to show the land-use classifications, which can also inform the developer in the conversations with the community."

The program is also planning 3D enhancements to the platform to make it easier for users to visualize the areas under consideration for development. "We're working on creating 3D map scenes

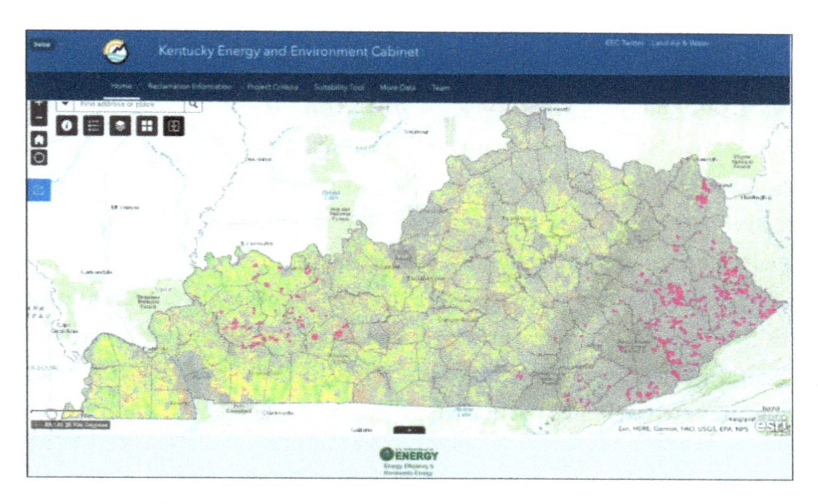

Interactive solar siting potential application created by the Kentucky Energy and Environment Cabinet.

in some of the prime areas for solar across the state," Miracle said. "This would allow developers to 'fly in' and see the slope, the terrain, a whole new view."

New applications, new opportunities

Kentucky's success in analyzing solar siting potential has created opportunities for collaboration with other states. "Given the work our GIS group is doing, we're known nationwide as the office you go to if you want to learn how to get started with GIS," Stump said. "We're answering energy questions from other states that, at the heart of it, are geospatial questions. Other energy offices are beginning to see the light: that they need to know where things are going to occur, where they should occur, and where they could occur, before talking to stakeholders and discussing policies."

Kentucky's GIS experts envision using the technology to adapt the siting platform tool to incorporate environmental justice data.

"It's going to be another lens by which we look at everything we do, from emergency response to permitting to siting facilities and economic development projects," said Stump. "Right now, we're assessing where the datasets are. We know the Environmental Protection Agency (EPA) has the EJScreen environmental justice tool, the Census Bureau has Community Resilience Estimates, the Department of Energy has the LEAD tool to examine low-income energy affordability, but how do we bring them together? What does environmental justice for energy look like in Kentucky? That's a big 'where' question. We're really excited about where we can go."

A version of this story by Mike Bialousz titled "Finding a Home for Solar: Kentucky Maps Prime Renewable Energy Sites" originally appeared on the *Esri Blog* on July 22, 2021.

WEB MAP BRINGS TOGETHER CONSERVATION AND GREEN ENERGY DEVELOPMENT

The Nature Conservancy

THE MIDWEST IS KNOWN AS THE WIND BELT OF THE UNITED States, and for good reason: nearly 80 percent of the country's current and planned wind energy capacity exists in the Great Plains, an area that extends east of the Rocky Mountains and runs from northern Montana to southern Texas. Wind energy shows tremendous potential as a clean, renewable energy source that can help reduce greenhouse gas emissions.

Much of wind energy development is occurring—and is expected to increase—in the wind belt. But as wind energy developers plan new sites, they face this question: How can new wind turbines be sited in places that are optimal for wind resources and transmission yet aren't likely to impact wildlife or encounter costly delays from regulatory or legal challenges?

Wind projects sited in the wrong place can threaten some of the best wildlife habitat. The Nature Conservancy (TNC) estimates that renewable energy development could adversely affect as much as 76 million acres of land in the United States—an area about the size of Arizona.

But a new GIS-based resource developed by TNC can help focus renewable energy in the right places—windy areas that pose a relatively low risk to wildlife and their habitats. Called *Site Wind Right*, this interactive online map is available for wind developers, power purchasers, utilities, companies, state agencies, and municipalities to help reduce conflict between wind energy and conservation.

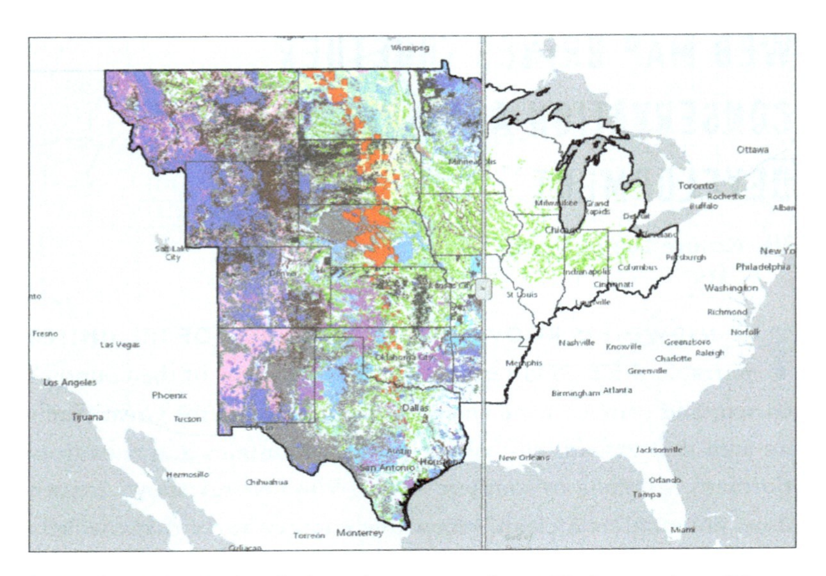

TNC's *Site Wind Right*, which evaluates more than 100 datasets from 17 states, shows that 90 million acres in the US wind belt could be developed for wind energy without affecting key wildlife habitats, which are depicted on the map by the varying colors. Map courtesy of TNC.

TNC developed *Site Wind Right* for 17 states in the Midwest, pulling from more than 100 datasets on wildlife habitat and land use to help highlight areas with the lowest potential for environmental friction. The results of this analysis, done by TNC scientists, are both enlightening and encouraging.

"We were thrilled to discover we could generate more than 1,000 gigawatts of wind power in the central [United States], solely from new projects sited away from important wildlife areas," said Mike Fuhr, state director of TNC in Oklahoma. "That's a lot of potential energy, comparable to total US electric generation from all sources today. While advancements in transmission and storage would be needed to fully realize this wind energy potential, it proves we can have both clean power and the land and wildlife we love."

Great potential for wind in the Great Plains

What eventually became the *Site Wind Right* analysis started evolving in 2011 for two reasons. First, wind energy facilities began to operate across the Great Plains. Second, TNC and other scientific studies demonstrated considerable potential for wind and solar energy development in the western and central United States.

The Great Plains is home to the country's largest and most intact temperate grasslands, which are among the least protected habitats in the world. The last expanse of this once-extensive ecosystem is found in the Greater Flint Hills region of Kansas and Oklahoma. Poorly sited wind turbines in places such as Flint Hills seriously threaten wildlife that depends on this endangered and beautiful place that is home to bison, bald eagles, and the once-common greater prairie chicken.

But as studies demonstrated, the Great Plains could provide clean, renewable electricity that doesn't compromise wildlife habitat and other natural resources.

"Those studies showed very positive results that we can meet or exceed renewable energy goals by using sites that were previously disturbed or had relatively low conservation value," said Chris Hise, associate director of conservation for TNC in Oklahoma.

Ultimately, TNC scientists created a resource that energy planners could use early in the siting process to avoid impacting wildlife and delaying their projects. TNC is among many organizations that want properly sited wind, solar, and other renewable energy projects to succeed to meet the challenges posed by climate change.

With support from partner organizations and other TNC scientists, Hise and his team collected vast amounts of data—on wildlife, habitats, land-use restrictions, areas of significant biodiversity, and more—and organized it in ArcGIS Desktop using ArcCatalog™. With ArcMap™ and ModelBuilder™, the TNC team then assembled

The temperate grasslands of the Great Plains—home to bison and other wildlife—are among the world's most altered and least protected habitats. Photo courtesy of Chris Helzer, TNC.

multiple spatial data layers of wildlife habitats and potential engineering and land-use constraints. Finally, using ArcGIS Web AppBuilder, the team created an online resource that could share this data in what became the *Site Wind Right* interactive map.

Hise and his team found an impressive number of low-impact areas across the central United States in the analysis—approximately 90 million acres. Planners in the early stages of establishing a wind energy operation can see site-specific details, explore *Site Wind Right*, consult with appropriate state wildlife agencies, and use the Wind Energy Guidelines developed by the US Fish and Wildlife Service to find spots that work best for everyone. And the low-impact sites in the Midwest are very well distributed.

"If we plan carefully, there's plenty of space to go big on wind energy in this part of the country," said Hise.

The once-common greater prairie chicken population has fared poorly as its grassland habitats have been converted to other uses. Photo courtesy of Harvey Payne, TNC.

Broadening the reach of wildlife-minded green energy projects

Site Wind Right has the potential to reduce the risks of wind deployment delays and cost overruns by helping developers locate sites that are less likely to face regulatory or legal challenges. This has spurred the endorsement of Evergy, an energy provider in Kansas and Missouri that became an early user of the analysis.

"*Site Wind Right* is an invaluable resource that helps us avoid unnecessary impacts to the wildlife and iconic landscapes of the Great Plains while also allowing us to provide clean, low-carbon energy for our customers," said former Evergy CEO Terry Bassham.

The mapping analysis invited accolades from another early reviewer, the Association of Fish & Wildlife Agencies, which conferred its 2019 Climate Adaptation Leadership Award for Natural

Resources on *Site Wind Right*. Additionally, the web map has received endorsements from several conservation groups, including the National Wildlife Federation and the Natural Resources Defense Council (NRDC).

"We need more resources like this to speed up our move away from burning fossil fuels," said Katie Umekubo, a senior attorney at NRDC. "Well-sited wind energy allows us to meet our climate goals, advances conservation, and ensures that we avoid irreversible environmental impacts."

Currently, TNC is looking to broaden the reach of *Site Wind Right* within communities, companies, and government agencies so they can quickly apply this wildlife-minded strategy and get the blades turning on clean and homegrown energy in the Great Plains.

"The Nature Conservancy supports the rapid acceleration of renewable energy development in the United States to help reduce carbon pollution," said Fuhr. "We are looking forward to providing *Site Wind Right* to the people making important decisions about our nation's clean energy future."

A version of this story by Eric Aldrich titled "Web Map Brings Together Wildlife Conservation and Green Energy Development" originally appeared in the Fall 2020 issue of *ArcNews*.

HOW ADVANCED ANALYTICS FUELS RENEWABLE ENERGY

Renewable Energy Systems; Equinor

AS GLOBAL CONCERN ABOUT CLIMATE CHANGE GROWS, countries and companies worldwide remain committed to reducing carbon emissions. Energy leaders are rebalancing their portfolios to include renewable sources such as wind and solar, and pure play firms (companies that focus on a single product) continue to build their businesses on alternative sources of energy. Both use location technology to drive profitable investments in this new energy landscape by cross-referencing cost data with factors such as ocean depth, wind speeds, and sun exposure.

As more projects have been developed over time, more players have joined the market, driving down prices for producers and customers. Large established energy producers, such as Equinor of Norway, as well as firms such as Renewable Energy Systems (RES), are seeing surging customer demand for renewable energy.

Corporate customers are showing interest too, entering agreements directly with renewable power producers to meet sustainability pledges, raising their profiles among eco-aware consumers. "The big retail stores and web giants are signing more and more contracts, because it enhances their green credentials, and it's in their financial benefit to do so," said Andy Oliver, chief technology officer (CTO) at RES, which has agreements with Google, Microsoft, and other large companies to provide cleaner power. This trend is one aspect of a broader shift toward corporate social responsibility.

When doing good for the environment also means lower bills, so much the better. Lower cost is indeed now a major driver of renewables. "Not everyone takes the climate threat seriously," Oliver said.

"However, paying less for something is usually something that everyone can agree on." The pressure is on energy companies to find the most cost-efficient ways to produce renewable power so that everyone wins. For many energy producers, location intelligence, fueled by sophisticated analysis from GIS, is an essential means of ensuring they make the right development decisions.

Putting renewables on the map

RES bills itself as the world's largest independent renewable energy company, with a presence in North America, western and northern Europe, Türkiye, and Australia—10 countries in all. To date, RES has developed or constructed projects that produce about 17,000 megawatts, which would power a country the size of Poland. Most of that energy comes from wind because of its historically low cost. But, according to Oliver, solar is increasingly taking a larger share in most of the company's markets, as solar technology costs have fallen precipitously.

There is a lot of money at stake. Typical US wind and solar projects run from $100 million to $300 million. Counterintuitively, the cheapest piece of land might not translate into the most valuable site. Many more factors affect the decision of where to put a project.

Like many companies, RES relies on location intelligence across multiple phases of its business—from assessing the viability of a project site to making construction faster, safer, and more accurate.

The cutting edge of project assessment

RES makes siting decisions for renewable energy projects more quickly and accurately than in the past. It uses GIS technology to track highly granular location information for solar and wind projects (such as airspace considerations, terrain slope, wind speeds, floodplains, proximity to power lines, soil composition, and setbacks from existing infrastructure) in conjunction with cost data. GIS-generated location intelligence helps RES pinpoint the right site with the highest ROI.

RES recently completed a project to learn whether solar or wind is more competitive in each of its markets worldwide. The challenge: use GIS analysis to determine how much energy would have been produced by each source at more than half a million locations around the globe.

"It's basically a weather forecast that you run backwards in time to understand what would've happened in terms of either wind production or solar production," RES's Oliver said of the big data project.

The result, in Oliver's words: "We now have a map, for example, of the United States that tells us, for any point, is wind more competitive at that location or is solar more competitive at that location? Having all that information available at the very beginning of a project will end up saving us time and money later by avoiding investment in bad sites."

By using location technology, he said, developers can understand in advance how much energy a site might produce, and then work with counterparts in the construction group to determine the building costs. "All of that goes in a financial model—all those costs and energy inputs—and that tells you what the cost of power is going to be," Oliver said.

RES teams work together to review locations for potential projects. Project managers review maps of permitting requirements, such as protected-wildlife areas, and degree of landowner interest, while specialists take a deeper look from a technical standpoint.

A wind project, for example, requires engineers to map the elevation and slope of an area. If the slope is too steep, the top of the turbine will not be reachable by crane, making the site unworkable.

A committed innovator, RES is using drones, geospatial data, and smart construction equipment to save time and increase accuracy when digging, surveying sites down to the subcentimeter.

The team creates a digital model of the surface of the project from drone imagery, upon which it bases drawings for the project, Oliver said. "We combine these two things into a file that can be read by a modern dozer or excavator." For example, as an operator drives the bulldozer, "it's reading where the surface should be, and it's moving its bucket up and down or side to side to make the surface that was envisioned in the drawing."

When digging a foundation for a wind turbine, location awareness helps the excavator reach the exact target depth, so the bucket will scoop the right amount of earth each time. Oliver said RES can save up to 40 percent of the time it used to take to dig a foundation and can avoid costly mistakes such as overexcavation and the subsequent need to backfill.

Powering self-service maps

In another example, midstream and upstream oil and gas provider Equinor is expanding its portfolio into renewable energy—primarily offshore wind projects—in its transformation to a diversified energy company. Like RES, Equinor uses location intelligence driven by GIS to guide siting decisions.

"Not very long ago, if people wanted maps for their reports, they would have had to call the mapping department," said Henrik Hagness, leading engineer for mapping at Equinor. But now, many professionals can handle their own queries using tailored interactive maps, driving a major increase in productivity and data-based decisions.

In a world with constraints on resources and limited untouched nature, it is more essential than ever to select the best location to optimize energy output with minimal impact. Equinor business development professionals weigh this and many other variables when deciding where to put an offshore wind farm. Location intelligence reveals important factors in the physical environment (measurements relating to the water, wind, seabed, and geology), as well as the socioeconomic environment (country borders, existing infrastructure, pipelines, cables, habitat restrictions, and easements) and supply chain considerations, such as logistics and shipping constraints.

Each project involves many potential players, Hagness said, ranging from governments and authorities who may set terms, conditions, and constraints, to providers of engineering, fabrication, construction, and installation services. "With GIS, we try to give the business development managers spatial intelligence and interactive maps of where we [might] pursue investment" Hagness said, so they can navigate through all the information to arrive at the right decision. But, as at RES, selecting the location is just the beginning.

"Spatial intelligence is needed throughout [the process], from business development and siting, to moving ahead with a development project, to efficiently maintaining and inspecting the energy assets when in operation, and eventually removing the assets and bringing nature back to its original state," Hagness said.

Investment guidance faster

Wind farms typically have a useful life of 30 years, from inevitable wear and tear, so it is important to seize the right opportunity at the right time. And that means paying close attention to the cost of energy, projected revenue, and other factors.

If Equinor wants to develop a wind farm outside Boston, for example, it needs to know the market for purchasing the services needed to build it and the distances involved in manufacturing the infrastructure and sending it to Boston for assembly, Hagness said. It's important to look at the supply chain as a whole and then compare it to different potential locations.

Financial analysts used to spend weeks getting numbers for each potential project, then updating their models with newer and fresher data. They consider everything from ever-evolving market data to changing political situations.

"Now we can do in hours the same thing that we used to spend weeks on," Hagness said.

Once Hagness's colleagues crunch cost and project variables, they use GIS to produce color-coded interactive maps denoting areas of greatest opportunity.

An Equinor business development manager can look at the interactive map of a prospective wind turbine site and, in that one view, see all relevant data—including elevation, slope, wind speed, waves, currents, soil composition, bedrock depth, ports, harbors, borders, existing infrastructure, shipping routes, and levelized cost of energy. The business development manager can check a box next to any of those layers to turn it on or off. He or she can zoom in and out and measure the site's proximity to transportation corridors. This capability is a major advantage for a company that wants to make efficient, data-based decisions.

"I think it is fantastic that you have different layers of intelligence that you can pick and choose from," said Elena Farnè, technical director at Equinor who assesses business cases for prospective offshore wind projects. These maps "ensure that you are…getting the kind of granularity of information you would like, especially in the early screening phase."

Shared insight born of complex data

Another boon to business decisions is that responsive GIS maps make it easy for teams to take another look at areas they have previously put aside, Hagness said. "If you had a model from 2017, you could recalculate all this [for the present]. Then you have two datasets. And what you do is you just drag the curtain back and forth, and you can see the differences." So, if the price of steel has come down or interest rates are lower, the conditions might be right to do the project today: "You can use the slider to see an immediate comparison," he said.

GIS data provides the teams a benchmark across renewable projects, Farnè said. "You have an immediate impression of whether you're competitive and where you stand in terms of cost in relation to production."

The technology makes it easy to share complex information, including models and dashboards, without the need to get on a plane. "The beauty of it is that it's all from a desktop assessment," Farnè added. "In a research-constrained organization, you can acquire data, and you don't need to travel. It's just a data-sharing exercise in most cases with our GIS specialists who update and enable the interactive maps."

As the pressure to mitigate climate change builds by the day, location intelligence supports traditional energy companies and pure play renewable providers alike. But in pursuing renewables, neither can afford to deploy resources in a haphazard way. Location technology reduces the risk considerably. Oliver of RES goes so far as to say, "Without good wind and solar sites that we've found through using GIS software, we wouldn't have a business."

A version of this story by Alessandra Millican and Geoff Wade titled "How Advanced Analytics Is Fueling Renewable Energy" originally appeared in *WhereNext* on April 21, 2020.

AN INDUSTRY ON THE VERGE: GREEN HYDROGEN

Esri

AS NET-ZERO PLEDGES RISE TO THE TOP OF CORPORATE priority lists, industry leaders are setting their sights on clean-burning hydrogen as a new ally in the quest for decarbonization. Today, hydrogen comprises just a small fraction of global energy use. The location technology already used in adjacent industries could accelerate the efficiency needed for hydrogen's ramp-up.

The United States wants to expand the hydrogen industry. Along with other countries, it also hopes to ensure a cleaner industry, in the context that up to 95 percent of hydrogen is produced from fossil fuels.

Hydrogen—typically converted to energy through fuel-cell technology—could meet up to 24 percent of the world's energy needs by 2050, according to a 2020 report from BloombergNEF. But there is much to be done. "Green" hydrogen produced with renewable energy accounted for just 0.1 percent of global supply in 2021. And it's expensive, costing up to US$6 per kilogram versus $1–$3 per kilogram for conventional hydrogen. To increase sustainable production processes, the energy industry must quickly make green hydrogen production more cost efficient. Technologies such as virtual 3D models, analytics tools, and real-time dashboards—all forms of location intelligence—can provide the insight that energy companies need to reduce cost and increase production.

Which color is your hydrogen?

Not all hydrogen is the same. Although all kinds of hydrogen production can benefit from location intelligence, net-zero advocates favor green hydrogen for these reasons:

- **Gray hydrogen:** This production method is by far the most common today, deriving hydrogen from the burning of natural gas. This creates nine parts of carbon dioxide for each part of hydrogen.

- **Blue hydrogen:** Companies producing blue hydrogen also burn natural gas but use carbon capture technology to mitigate carbon dioxide emissions.

- **Green hydrogen:** The most sustainable method of producing hydrogen, this process uses energy from wind, solar, or another renewable source to split water into hydrogen and oxygen, producing no carbon emissions.

The hydrogen resulting from these processes is clean burning and often used in fuel cells to produce energy.

For hydrogen producers, the cost of going green is largely determined by one factor: a hydrogen plant's location. Green hydrogen can be expensive to produce through renewable energy if input costs are high, and it can be expensive to store and transport. Energy companies entering the green hydrogen market must first understand location-based variables such as regional access to renewables, local demand for hydrogen, and nearby infrastructure for hydrogen transport (for example, shipping versus pipelines).

These variables inform a series of decisions—whether to convert existing plants or construct new ones, which type of renewable energy should power the electrolysis process that produces hydrogen, and whether to store and distribute the hydrogen as a gas or liquid.

As leaders in green hydrogen consider where and how to optimize costs, they can learn from energy companies that have relied on location technology for years. The oil industry company, BP, for instance, uses GIS for everything from pipeline management to market research. In renewables, RES and Norway's Equinor use GIS for site selection and other critical processes. Hagness, Equinor's leading engineer for mapping, said spatial intelligence is "needed throughout [the process], from business development and siting to moving ahead with a development project to efficiently maintaining and inspecting the energy assets when in operation."

In growing industries such as green hydrogen, GIS-based smart maps and dashboards can provide single-pane-of-glass visualizations that identify cost-effective options and point toward evidence-based next steps.

Bringing data together drives smarter improvements

Once green hydrogen systems are in place, energy companies will need to stay agile, fine-tuning operations and applying efficient processes at scale. Reliable growth is best achieved by equipping executives with a decision support system—a holistic operating picture based on analytics that reveals patterns and outliers and models potential outcomes.

For green hydrogen producers, key performance indicators (KPIs) may include a plant's renewable energy consumption rates, local weather patterns, buyer demand, and equipment conditions. For executive-level decisions, these datasets are best viewed together, rather than siloed. For example, comparing weather patterns and buyer demand—an analysis that natural gas operator ONEOK Inc. performs with GIS—enables more nuanced decisions about optimal times to run electrolysis.

Location analytics tools combine these pieces of information and identify key connections. Visualizing the data on map-based dashboards shows where even small operational adjustments will have the most impact. Green hydrogen companies may find inspiration in energy firms that have used decision-support technologies such as GIS to integrate analytics and visualization and ensure that executives have access to complete, current information.

Looking ahead with real-time awareness

As the green hydrogen industry expands, the demand for day-to-day operational efficiency will increase. Keeping track of equipment maintenance, clean energy certifications, and safety protocols will require a location-based view of assets and activities. For some hydrogen producers, IoT sensors and real-time data feeds will supply accurate and near real-time information on those conditions and help avoid costly downtime.

Combining that data with maps and dashboards for visualization, companies can even create a digital twin—a virtual model of a facility or supply chain that links information to location, showing executives what's happening where at any moment.

World leaders have indicated they will continue to prioritize green hydrogen and other clean energies in decarbonization efforts. Across the energy sector, technology that delivers intuitive access to operational information is already considered an essential business system. Leaders can turn to maps, dashboards, and digital twins populated by real-time data to manage the production of clean energy for a sustainable future.

A version of this story by Alessandra Millican and Geoff Wade originally appeared in *WhereNext* on August 17, 2021.

NEXT STEPS

The Geographic Approach to natural resources

NATURAL RESOURCE ORGANIZATIONS USE LOCATION-based data to find, share, and analyze information when they need it. GIS provides access to global spatial information, streamlines daily workflows, helps uncover trends, and improves understanding and decision-making. Here's how you can get started.

Implement your GIS

If your organization is not already using GIS, several resources are available to help you get started.

ArcGIS Online

With a subscription to ArcGIS Online, organizations can manage their geographic content in a secure, cloud-based environment. Members of the organization can use maps to explore data, create and share maps and apps, and publish their data as hosted web layers. Administrators can customize the website, invite and add members to the organization, and manage resources.

- ***ArcGIS Online Implementation Guide:*** Learn essential tasks and best practices for setting up ArcGIS Online.

ArcGIS Pro

ArcGIS Pro is the latest professional desktop GIS application from Esri. With ArcGIS Pro, you can explore, visualize, and analyze data; create 2D maps and 3D scenes; and share your work to your ArcGIS Online or ArcGIS Enterprise portal.

- ***ArcGIS Pro Implementation Guide:*** Learn essential tasks for getting your organization started with ArcGIS Pro.

Identify foundational data

Gather and map your foundational data for areas covered by your organization. Layers, such as the ones included in this list, may include basic infrastructure and administrative areas, as well as information important to your field of study:

- Administrative and jurisdictional boundaries (city and country boundaries, and so on)
- Population and demographics
- Infrastructure (roads, bridges, dams, utilities, communications, and so on)
- Major facilities and landmarks
- Water features (lakes, streams, rivers, and so on)
- Parcels
- Addresses
- Forest or field boundaries
- Rights-of-way

Add ready-to-use, curated content from ArcGIS Living Atlas of the World, which contains several live feeds that provide dynamic, real-time information in addition to your local data:

- Weather feeds
- Earth-observation feeds
- Multispectral imagery feeds

Discover more live feeds in ArcGIS Living Atlas at links.esri.com/atlas_live.

Also consider adding real-time services for additional situational awareness:

- Operational data
- Worker locations
- Sensor readings

Learn by doing

Hands-on learning will strengthen your understanding of GIS and the ways you can use it to improve organizational efficiency and effectiveness. ArcGIS documentation includes a collection of free online story-driven tutorials that allow you to experience GIS when it is applied to real-life problems. For the gallery of tutorials, go to product documentation at https://doc.arcgis.com.

Learn the basics

Product documentation includes these and other tutorials to get you started on your GIS journey:

- **Create a map:** Create a web map in ArcGIS Online.

- **Create an app:** Configure and share an app that puts your web map to greater use.
- **Get started with ArcGIS Online:** An introduction to web mapping using ArcGIS Online.
- **Get started with ArcGIS Pro:** Learn the basics of ArcGIS Pro.

Natural resource-specific tutorials

Product documentation also includes these and other tutorials applicable to natural resources:

- **Plan a timber harvest:** Locate and measure a proposed timber harvest.
- **Manage forest harvest and chemical activities:** Use the forestry activities feature layer template to capture harvest block locations, associated forest data, and track treatment activities.
- **Perform a site suitability analysis for a new wind farm:** Determine the optimal location for a set of new high-efficiency wind turbines in Colorado.
- **Monitor forest change over time:** Detect and analyze forest disturbance and recovery from a Landsat time series in the West Cascades ecoregion in Oregon.
- **Get started with multidimensional multispectral imagery:** Use a multidimensional stack of Landsat imagery to visualize how a Chilean copper mine has changed over time.
- **Explore and animate geological data with voxels:** View and analyze multidimensional soil voxels for the Netherlands.

- **Assess hail damage to cornfields with satellite imagery in ArcGIS Pro:** Compute the change in vegetation before and after a hailstorm with the Landsat Soil Adjusted Vegetation Index (SAVI).
- **Use deep learning to assess palm tree health:** Identify trees on a plantation and measure their health using imagery.

Get there faster with GIS templates

ArcGIS Solutions for natural resources streamlines deployment of location-based solutions in your organization. You can use these solutions, or templates, to take action that helps you understand your business and manage operations.

ArcGIS Solutions for natural resources includes these templates and more:

- **Market Development:** Better understand market characteristics and performance. It includes a Market Explorer Dashboard and a Sales Performance Dashboard.
- **Business Resilience:** Better understand the status of asset locations. This solution provides key leadership the ability to share the status of retail stores, manufacturing plants, and other facilities.
- **Incident Analysis:** Identify patterns and trends of significant activity. This solution contains maps and tools to perform pattern and trend analysis and allows you to author situation maps showing incident locations and types of activity.

- **Fusion Center:** Conduct threat mitigation on incidents affecting organizational assets from multiple event feeds. This solution delivers a set of capabilities to correlate information from multiple event layers and assess the impact to an organization's assets, such as facilities and employee home locations.

- **Gas and Pipeline Referencing Utility Network Foundation:** Accelerate a unified utility network and Pipeline Referencing implementation for gas and hazardous liquid pipelines. Starting with this foundation frees organizations from spending their time and implementation budget on building custom data models and information products.

Learn more

For additional resources and links to live examples, visit the book web page:

go.esri.com/mow-resources

CONTRIBUTORS

Matt Ball
Jim Baumann
Chris Chiappinelli
Mark Dann
Matthew Harman
Charlie Magruder
Sarah Powell
Monica Pratt
Ben Smith
Citabria Stevens
Lo-Ruhama Westbrook
Carla Wheeler

ABOUT ESRI PRESS

AT ESRI PRESS, OUR MISSION IS TO INFORM, INSPIRE, AND teach professionals, students, educators, and the public about GIS by developing print and digital publications. Our goal is to increase the adoption of ArcGIS and to support the vision and brand of Esri. We strive to be the leader in publishing great GIS books, and we are dedicated to improving the work and lives of our global community of users, authors, and colleagues.

Acquisitions

Stacy Krieg
Claudia Naber
Alycia Tornetta
Craig Carpenter
Jenefer Shute

Editorial

Carolyn Schatz
Mark Henry
David Oberman

Production

Monica McGregor
Victoria Roberts

Sales & Marketing

Eric Kettunen
Sasha Gallardo
Beth Bauler

Contributors

Christian Harder
Matt Artz
Keith Mann

Business

Catherine Ortiz
Jon Carter
Jason Childs

For information on Esri Press books and resources, visit our website at esri.com/en-us/esri-press.